# "The Horror of the Fight"

## The Civil War Diary of George Fairhead

Edited by

James S. Pula

ISBN: 978-0-941504-00-3

WinStan Press
Academic Consulting
Washington, DC



The editor wishes to thank the Oneida County History Center,
Utica, New York, and its director, Brian Howard,
for permission to publish the Fairhead diary and accompanying images.

To the Memory of My Schoolmate,
Collaborator, and Friend

Eugene E. Dziedzic

1862  1899



*The Thirty-seventh Annual Reunion*

*of the*

*117th N. Y. Vol. Veteran Association,*

*will be held at*

*Summit Park, near Oriskany, N. Y.,*

*Tuesday, August 22,*

*1899.*

(OVER.)

# Contents

Printed muster roll of the original Company D, 117th New York. Fairhead's name appears under the heading "Sergeants" as the second listing in the column on the right.

*Oneida County History Center.*

# Introduction

Born in Norwich, England, on January 5, 1835, George Fairhead came to the United States with his mother and three siblings, arriving in New York aboard the ship *Gladiator* on December 31, 1840. Educated in the local public school in New York Mills, he worked in the textile mills there from 1845 to 1854 after which he obtained a job clerking in a local business before entering the Whitestown Seminary to study for the ministry. The onset of the Civil War found him as both a student and also apparently teaching at least part time. When New York Governor Edwin D. Morgan called for the recruiting of two new regiments from Oneida County in July of 1862, Fairhead responded, mustering in as a corporal in Company D, 117th New York Volunteer Infantry, on July 25. The enlisting officer recorded him as 27 years old, 5'6" in height, with a light complexion, blue eyes, and dark hair. He listed Fairhead's occupation as "student."[1]

Assigned to the defenses of the nation's capital, Fairhead's regiment spent the remainder of the year digging fortifications near Tenallytown (now Tenleytown) on the western edge of the District of Columbia. In March, 1863, when his application for a brief leave gained approval, he hurried home to be married to Hannah Ackroyd. Also a native of England, Ackroyd arrived in New York with her mother aboard the ship *Sheridan* at age two on January 1, 1842. The couple knew each other growing up in New York Mills, but traveled to Holyoke, Massachusetts, to be married at the home of Hannah's older sister. Fairhead traveled there for the wedding, then escorted his new bride back to New York Mills by train. Three days later he left to rejoin his regiment, not to see his wife for another year and a half.[2]

Fairhead no sooner arrived in Washington than his regiment was transferred to Virginia to defend Suffolk from an advance by Confederates under Gen. James Longstreet. After participating in a march toward Richmond, the War Department assigned the 117th to the forces besieging Charleston, South Carolina where it served until April 1864 when it was assigned to Gen. Benjamin Butler's Army of the James operating

---

[1] Manifest of the ship *Gladiator*, Ellis Island ship manifest collection; George Fairhead Folder, OCHS; Fairhead annotated roster; Regimental Descriptive Book Companies A-K, 117th New York Infantry, RG 94, National Archives and Records Administration. A memorial card in the collection says that he weighed 2 ½ pounds when born and was "frail."

[2] Manifest of the ship *Sheridan*, Ellis Island ship manifest collection; George Fairhead Folder, OCHS; "Fifty Years Married," unidentified newspaper clipping, Oneida County History Center.

against Richmond and Petersburg in Virginia. There, it fought at Bermuda Hundred, Swift Creek, Drewry's Bluff, Cold Harbor, the Siege of Petersburg, Chaffin's Farm, Fort Harrison, Fort Gilmer, and the Darbytown Road. Transferred again, it led the assault that captured Fort Fisher, North Carolina, in January 1865, then marched through Wilmington to Raleigh to join Gen. William Tecumseh Sherman's forces moving north through the Carolinas. Promoted to sergeant, Fairhead was in the thick of the fighting, sustaining a serious wound at the Darbytown Road. After recovering, he gained promotion to sergeant major and, in late March of 1865, received a commission as second lieutenant from New York's Governor Reuben Fenton.[1]

After mustering out of the service in June 1865, Fairhead returned home to Hannah. The couple had two sons. Fairhead taught for three years before entering the Methodist Episcopal Ministry in April 1868. Over the succeeding years he served congregations in Steuben, Brookfield, Norway, Springfield, Starkwell, Munnsville, Mohawk, McComb, Floyd, and Hamiline before teaching for five years at the Cazenovia Seminary. A staunch Republican, he became very active in veterans' affairs in the Grand Army of the Republic, joining Ross Post 31 in New York Mills where he served as commander in 1905 and 1906. On the state level, he gained election as chaplain for the Department of New York in 1894 and edited the *Grand Army Journal*, a weekly newspaper "for the surviving soldiers and sailors of the war for the union," for eight years from 1886 to 1894. In 1934 he attended the G.A.R. encampment in Rochester, N.Y., at the age of 99 years and 7 months, the oldest person there.

Fairhead spent the final years of his life at the Utica City Home for Aged Men and Couples, a section of Faxton Hospital. He passed away on October 23, 1939, at the age of 104.[2]

▪ ▪ ▪ ▪ ▪

"Not known to the historic pen, or platform orator," Fairhead wrote on the first page of his diary, "but the private led in the horror of the

---

1 Frederick Phisterer, *New York in the War of the Rebellion* (Albany: J. B. Lyon Company, 1912); *The Union Army: A History of Military Affairs in the Loyal States, 1861 -65 – Records of the Regiments in the Union Army – Cyclopedia of Battles – Memoirs of Commanders and Soldiers* (Madison, WI: Federal Pub. Co., 1908, Vol. II.

2 George Fairhead Folder, OCHS; letters, Alan P. Davis, President of Rochester Historical Society to Geoffrey C. Clifford, Curator, Oneida County Historical Society, April 5 and May 2, 1966, Manuscript Box 4, 1986, MSS.1, FAI.1, OCHS.

George and Hannah (Ackroyd) Fairhead in later years. *Oneida County History Center.*

fight all the same."[1] The diary he kept for the last fourteen months of his service provides a lens through which one can view the war as it appeared to the average soldier—the monotony of camp and drill, the exhaustion of long marches, the every-day routine of life with ones comrades, each punctuated by the sheer terror of the battlefield. All of this is captured in direct, sometimes cryptic comments written at the time the events occurred. Because of this, the diary reflects not only his impressions and anecdotes of his experiences and those of his friends, but also what an average soldier believed to be important. We see first-hand the privations, constant tension in the front-line trenches, and the sorrow over lost friends. His constant reference to letters sent and received clearly demonstrates the importance a soldier far from home placed on keeping in touch with distant friends and loved ones.

Aside from illuminating the human aspect of the Civil War, Fairhead's observations provide valuable information for historians because the 117th New York, known locally as the "Fourth Oneida," participated mostly in the campaigns of the Army of the James and Fort Fisher, both of which have been studied substantially less than those of the Army of the Potomac or the campaigns of Grant and Sherman in the west.

[1] George Fairhead diary, Oneida County History Center, Utica, New York.

▪ ▪ ▪ ▪ ▪

In the diary, original capitalization, punctuation, and spelling have been retained.

The editor wishes to express his appreciation to Brian Howard and the Oneida County History Center for permission to publish the Fairhead diaries and illustrations from its archives; Donald Wisnoski for permission to use images from his vast collection of Civil War photographs; and Hal Jespersen for masterly converting my rough sketches into the exquisite maps that appear in this volume.

# Drewry's Bluff

## May 1864

*George Fairhead's diary begins in May, 1864, on the eve of the Battle of Drewry's Bluff. The engagement resulted from Gen. Ulysses S. Grant's belief that to defeat the Confederacy he must take advantage of Union numerical superiority by placing all of the Federal armies in motion at once to prevent the Rebels from shifting men to meet new threats. The Army of the Potomac under Maj. Gen. George Gordon Meade, with Grant accompanying it, was to move south attempting to bring Gen. Robert E. Lee's Army of Northern Virginia to a stand in the open where it could be defeated. Maj. Gen. William Tecumseh Sherman would lead an army south from Chattanooga through northwestern Georgia aiming toward Atlanta, destroying railroads and Southern supplies along the way while attempting to neutralize the Confederate army initially led by Gen. Joseph E. Johnston. In Virginia, Grant ordered Maj. Gen. Benjamin Butler's Army of the James to proceed up its namesake river to attack Richmond. The purpose of Butler's movement was to prevent the Confederates from reinforcing Lee's army, giving Meade a better opportunity for success. Other smaller Union forces were to make whatever advances they could to further tie down Rebel troops or eliminate railroads, warehouses, manufacturing establishments, or other support for the Confederate war effort.*

*Assigned to Butler's army, the 117th New York was part of Col. Samuel M. Alford's brigade including the 40th Massachusetts and the 3rd, 89th, and 142nd New York. Their mission was to threaten Petersburg and Richmond to pin down Confederate troops, and if practical to assault either city. The movement began with the receipt of orders on April 29. That evening the regiment marched to a landing place where it boarded the steamer* Columbia *at 4:30 am to carry it to Bermuda Hundred, a small peninsula located at the confluence of the James and Appomattox Rivers south of Richmond. The troops landed on May 6 and Fairhead began his journal eight days later as they were advancing on the Confederate fortifications centered on Fort Darling at Drewry's Bluff about twelve miles south of Richmond.*

---

George Fairhead in 1863.
*Oneida County History Center.*

May 14th Saturday. In battle line near Fort Darling, Va. Rain with intervening intervals of sunshine. At Early morning our forces changed line by filing through the woods going to the position held by the rebs yesterday. They had vacated during the night. The Position was a very large cleared field having rifle pits around the whole, battery with Embrasures at one side. We passed over this, & formed new line. This line faced other rebel works & ran along a road. The Skirmishers were in advance. Bullets from the rebs would frequently "flit" by. Laid down along this position. Remained all day. This place Shot & shell flew over us all day long. Engagement between Skirmishers along the whole line. The reb bat[tery] was sharply engaged by our bat[terie]s. The rebs ventured to change some of their artillery somewhat to the left of the line. A new battery opened on them some what suddenly. Our forces then having a cross fire on the rebs who scrached for their earthworks in haste. A charge was made late in the after part of the p.m. to take the reb bat[tery]. Our men got within speaking distance of the fort. Laid there a while when the rebs made a sallie[1] over their parapets. They were many. Bullets flew like hail. 4 of our co[mpanie]s were in the charge. Our men laid behind a fence. The fence was riddled. Our men finding it out of practicability to make the sacrifice to force their way through, fell back after the rebs had fired their fill. A few of our men (only) were hurt, which is wonderful. A man in the 3rd N.Y. who had been at the front nearly all day was shot as he was passing to the rear. He had been talking with some of our boys, & had just past me, a bullet whizzed by, I heard it, twas near. He jumped some what lively, saying "I am shot." I though he was sporting. He went a little farther, & died. The bullet hit him in the back passing through his bowels. This was after dark. Strange that he should be engaged all day, & though leaving the Picket line in safety, a bullet should

[1] Sally: A sudden rushing forward of troops, especially from those besieged or holding a defensive line.

← Lt. Col. Rufus Daggett, *Oneida County History Center*

Capt. Egbert Bagg → *Donald Wisnoski*

reach him so fatally. Our men fell back to the Battleline. Laid all night, quite quiet.

May 15th Sunday. In battle line near Fort Darling, Va. Some cloudy, wind south of east. In same place. Rebs quiet. Our skirmishers are firing at rebs who are digging & throwing up earthworks. Skirmishing continued all day. Considerable quiet. 117th in Battle line near Fort Darling, Va.

May 16th Monday. In battle line near Fort Darling, Va. Heavy fog at Morning & till 10 a.m. Rebs got reinforcements during the night. Under cover of fog they massed them on the right of line which was pressed desperately. Our forces began a retrograde movement during last night. The engagement on right was terrific. The Rebs charged many times, were severly repulsed. Our troops were maneuvering to fall back, & fighting at the same time. Our Division was last, this Brigade leaving the last. 117th in line behind the 142nd, laying on our bellies. Shot, shell, shrapnel, canister and bullets whistled over us, a few men were wounded & killed in this position. The 142nd broke up & were rushing back on us. They were rallied, & gave the Rebs a heavy vollie. Dagget & Capt. Bagg[1] aided materially in rallying the 142nd. 117th got orders to move behind the Parapet, face old reb works, we did so. Got partly

---

[1] Lt. Col. Rufus Daggett, a carpenter from Utica, later became colonel, rose to brigade command, and received a brevet promotion to brigadier general. Captain Egbert Bagg, a Utica surveyor, later became major, commanded the regiment in four engagements, and was breveted lieutenant colonel. Brevet promotions were given for gallant and meritorious conduct in an era when the use of medals was uncommon.

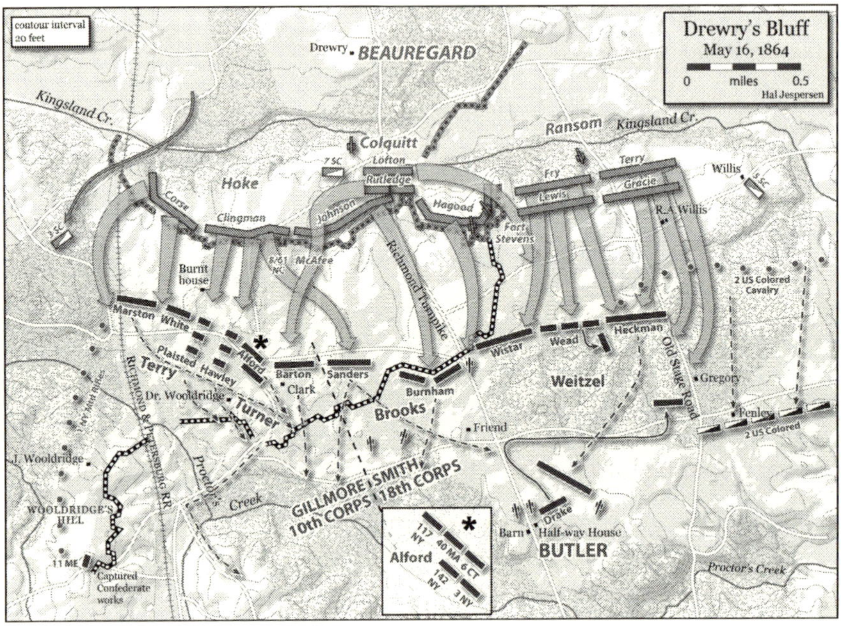


The 117th New York served with Alford's brigade at Drewry's Bluff. The cutout shows the alignment after the regiment came to the aid of the first line. *Hal Jespersen.*

---

separated. Part of the regt were on the slope & ditch side of the work, while the other part were on the inner or face side. The 142nd and 40th Mass. left their position. The 117th now moved in considerable broken order away from the work. The 142nd rushed into us which did not help us in the least. The Brigade was in a bad shape. The Rebs, coming on by the flank, going [toward] the flank of their old works. All sorts of deadly missels were screaching across the field, wounding & killing. To check the Rebs & show them a front while the Brigade was getting away, was the object now. Brig. Gen. [John Wesley] Turner gave us (117th) orders to form just behind the Brow of the hill, or rolling swell of ground. We formed on the colors, making a passible line. 117th charged forward to the top of the rolling sweep of land and gave the rebs a volley. The Rebs halted & fell behind the old work. 117th laid down on their bellies & continued firing some time, then moved back & formed new line near a fence & wood at the foot of the sweep of land. The Brigade now moved away. The 117th sustained a sever[e] loss in this movement. The broken order in which we fell back saved many a man from injury. While we (117th) were moving rapidly about & charging to the brow of the hill, (The men rushed in with almost wild enthusiasm) the other regts were busy in making new lines. The 40th Mass. moved in good order, not coming near the 142nd & 117th but passed to the rear on our right, or more strictly left of the battle line. They took position behind a second fortification & sent out skirmishers who held their place till the troops had got in to a hollow & moving to a second hill. Our men brought away most of the wounded. The dead we could not move (sad). The rebs came no farther in this direction at this time. A Battery of ours was in place on the hill to which we moved. We now fell back to the rear & to the turnpike, thence to camp, reaching which about 7:30 p.m. Report says that the Rebs shot any of the few wounded who were left behind, who could not walk. The Rebs had skirmishers dressed in U.S. uniforms. I saw the Skirmish line composed of them dressed in this manner. This deceived many of us & did much to get us into such a miserable fix. The 117th show good fighting qualities somewhat to[o] noisy which will be counteracted. Brig Genl Turner said "The 117th will go where it is sent" but they are to[o] noisy. Men in the ranks would repeat the commands, this produced the undue noise. Our Boys rally on the colors with the utmost ardor. The injury to the regt during the five days were Killed 16[,] wounded 65 and missing 7 [for a total of] 88.[1]

May 17th Tuesday. In camp in the field a few miles from Bermuda Hundred, James river, Va. Laid in camp all day. Pleasant weather. Passing clouds. Washed myself thoroughly, washed shirts, socks, towel and Handkerchief. Wrote to wife, have recd no letter from her since the 21st of April. Wrote for S. Sharp, wrote a note home.

May 18th Wednesday. In camp. Picket firing at the front during night somewhat brisk, but it ceased. A detail of 200 called out of the regt at 2 a.m. this morn. In camp till near noon. General Inspection ordered at 11 a.m. Orders came to move toward the fortifications. Rebs making some demonstration. They are getting a wagon Supply train through on the Turnpike. Our cavalry made demonstrations or sallies to prevent. The rebs are exceedingly vigilant and active in pushing this train onward. The Cavalry failed to capture the train. Came back to our camp at 8

---

[1] The regimental morning report for May 17 listed 16 killed, one mortally wounded, 34 severely wounded, 30 slightly wounded, and eight missing, a total of 89. *War of the Rebellion: A Compilation of the Official Records of the Union and Confederate Armies* (Washington, DC: U. S. Government Printing Office, 1880-1901), Vol. 36, Pt. 1, 21.

o'clock p.m. Read news after "Tattoo"[1] then went to bed.

May 19th Thursday. In camp in the field. Rainy & [lousy?] still. Roads in miserable order. Called out at 2 a.m. Moved to the position where we laid yester afternoon. Remained till 12 m. Returned to camp. Rebs cannonaded our locality considerable. Throwed many shot & shell. They did little or no damage. Our guns did not reply at all. If they charge our fortifications as we often (& nearly always), do theirs, they will find more than a volley of musketry. Genl Inspection this p.m. Shepard Inspected. All quiet at sundown. The Gun boats have been shelling something most of the day. Went to bed 9 O'Clock.

May 20th Friday. Bivouacked in field. Rain mostly over. Called out under arms at 12 Midnight, formed line, stacked arms, laid down, slept till Early morning. Picket firing was rapid at intervals. At early morn moved to an open field just behind the Batteries. During the morning the Rebs made an advance. The cannonading & musketry was to[o] much for them. Our batteries are terriffic when they open their blazoned throats. Laid in the field all day. Picket firing. Rebs made no more assaults to day. Sheltered ourselves from the sun by putting up our rubber blankets, or making Bivouacs. All troops are under arms. Those engaged on the fortifications (shoveling or chipping) have their arms stacked near them. Men in excellent spirits. They don't like to see men fall, fighting is not pleasant, but not any show the 1st sign of shirking duty. The dreadful work of battle must be done, & when the contest comes, each will join in the grapple. Slept on the field. Rubber blankets & blouse is all the bed we have & we enjoy even this. Called up at 2 a.m. Went to the Batteries & constructed abattis[2] in part of the lines. There were 7 co[mpanie]s out all night. They were constructing a strong fort on a hill some little in advance of the main line.

Got letters from wife (2) the first in a month. Read them by the light of a bivouac fire kept burning by fence rails. Letters from home exhilerate the feelings of soldiers to a high degree of vivacity. I felt firmer by reading. Very quiet during the night.

May 21st Saturday. Bivouacked in field. Sun unclouded. 117th quit work & took our station in the field (7 a.m.). The Rebs at work on a

---

1 Tattoo: A bugle call signaling "lights out," the time to retire for the evening. It preceded "Taps" which was coming into use at the time.

2 Abatis: A field obstruction made of sharpened tree branches laid in a row with the points facing toward the enemy.

This photograph, taken along the Petersburg fortifications, shows trees that have been cut and placed facing the enemy as a defensive abatis to impede any troops attempting an advance. *Library of Congress.*

battery. Our Bat[terie]s opened on them. Some Picket firing during day, laid on the field all day. Got soft bread, the first we have had since leaving Folly Island, S.C.[1] Men were as pleased with it as a boy with new boots. Our Hardtack[2] have been excellent. Soft bread is a treat. Rumor about in the early evening that the Reb Genl had sent in a flag of truce with the message to Our Commander to leave the position within 7 hours. Every thing was as quiet along the lines during the Early night as thoug[h] was not so near. A solitary pop of a musket was all that disturbed the stillness. A Calm. At 11:00 p.m. a musket or two, a volley, many vollies. All were waked, under arms in an instant. The Batteries began their terrible work. The showers of shot, shell, grape, canister & bullets which were sent at the rebs were to[o] dreadfully scathing for any number of men to face and live. The Brigade was not moved, those

[1] The regiment had been assigned to the siege of Charleston before moving to Virginia.

[2] Soldiers differentiated between "soft bread," the normal fare of civilians, and "hardtack" issued to the army. Hardtack was a very hard, dry cracker that was often softened with grease or by dipping in coffee before eating.

in front being sufficient. The Rebs recoiled how Severely they were punished we probably shall never know. This bellowing, screeming, death dealing storm lasted only 40 minutes. And again all was as silent as the grave. Except the hum of voices of the men as they again composed themselves for another nap. A caisson or two of the rebs blew up. The sky was lurid with the flash. The moon was 1 ½ hours high. All quiet the remainder of the night. Our Picket line was not driven.

May 22 Sunday. Bivouacked in the field, Bermuda Hundred, Va. Warm. Our line of batteries & Pickets are quiet to day. Our losses were very small last night, only 3 or 4 of whom I can learn. The enemy used very little artillery. A shell or two came into the field where we were but did no damage except to kill one horse. I think we shall take our time about leaving this position. Maj. Genl. Walker (reb) was taken prisoner on the 20th.[1] He refused to surrender at first, bullets brought him to terms. Shot in legs & arm. The Gun boats are sending shot at something (slowly). Other report say[s] that a shell hit him. His leg was awful shattered. Horse killed. At 4 p.m. we left the field in which we had been resting during the past 3 days. Marched back a little distance, & taking position in a forest. The underbrush had been cut & cleared away by the Pioneer Corps. Stacked Arms. Went to old camp, got tents, returned & put them up. Made brouse beds.[2] Went to bed at 9 p.m. At 11 pm. our regt & 40th Mass were ordered to fortifications. Went and relieved the force there. Laid down on arms ready for action.

May 23rd Monday. In new camp in forest. Excessively hot. Came to camp at 9 a.m. quiet during night, rearanged camp to day. Only a stray musket shot, & a few cannon have releaved or broken the quiet to day.

Wrote to wife. Had some talk with Br. Bushnell of "B" co.[3] Subject Christian Holiness. Was at my tent till "Tattoo." I went to bed immediately after. A beautiful bed composed of the splendid sprigs of the Oak Pine & ash, with a sheet of rubber blanket covering overcoat.

May 24th Tuesday. Camped in forest, Va. Air red hot. A detail of 260 men went (building road) for fatigue. Others remained in camp all day.

---

1 Maj. Gen. William S. Walker of Florida had his horse killed and his foot so severely wounded it had to be amputated.

2 A browse bed was one made in camp from the boughs of balsam, hemlock, spruce, or other soft wood.

3 James H. Bushnell, a Utica mechanic, was wounded at Fort Gilmer on Sept. 29, 1864, and died on Oct. 14 at a hospital in Hampton, VA.

Some few cannon shot somewhat rapid firing before the Pickets at early morn. We were got out under arms at 4 a.m. Remained all daylight. Read news & other things to day. Sung [?] somewhat in a small Patriotic note book I picked up on a field. Wrote to Sister Annie.

May 24th Wednesday [actually 25th]. Camped in forest, Va. Hot, Thunder clouds were rolling through the sky most of the day. 300 men were out on fatigue. I was out from Co. "D." I was with the men who were continuing at work on the road. A heavy storm came up just as we finished work. We were a mile or more from camp. Became soaked nicely before we got to our tents. Many had not taken their Rubbers.[1] I was in this predicament. As usual we were waked at 4 a.m.

May 26th Thursday. Camped in forest, Va. Cloudy & misty at Morn. Hot at M[noon]. Heavy shower at 4 p.m. 300 men worked on the new fortifications at the front. Ceased at 4 p.m. Nearly all of us had rubbers got but slightly wet about the shins. Roused at 4 a.m. An occasional musket shot from the from the enemy. Our Batteries sent a few shells over to greet the Grays.

May 27th Friday. Camped in forest, Va. Warm with rolling clouds. In camp all day till we changed locality & encampment. Wrote to wife. Also wrote to my father. Broke camp in the forest. All expected we were to move some distance. Went only about a mile.

---

*In its first major engagement, the 117th New York acquitted itself exceptionally well. During the Union advance it steadily moved forward in the face of enemy fire. When its brigade withdrew, it made a determined stand as the rear guard, saving the 40th Massachusetts from encirclement and covering the withdrawal of the rest of the brigade, the last unit of its division to leave the field. Following the action the colonels of the 142nd New York and 6th Connecticut each affirmed that had the 117th not held firm their regiments would have been cut off and probably captured. After withdrawing from the field, Massachusetts Col. Guy Henry "had his men in line beside the road, and as we*

---

[1] A reference to the rubber blankets the soldiers carried.

*came by had them present arms, dip their colors, and give three cheers for the 117th N. Y. Vols. for saving them."*[1]

A dark blue forage cap with black visor and brass fittings worn by a member of the 117th New York. The "D" above the regimental number indicates this belonged to a soldier in Fairhead's Company D. *Oneida County History Center.*

---

[1] Letter, "Nemo" to the *Utica Morning Herald*, May 24, 1864; Milton H. Culver, *Sparks from the Camp Fire of the 117th N.Y. Volunteers, from 1862 to 1865, Relating a Few Historical, Truthful, Thrilling and Mirthful Incidents as Gleaned from the Diary of Sergeant M. H. Culver, of E Co.* (Rome: Guernsey, 1917), 14.

# Cold Harbor and Petersburg Heights

## May 29—June 17, 1864

*Following the clash at Drewry's Bluff, Gen. Butler retreated back into the defensive line at Bermuda Hundred. Meanwhile, Gen. Grant prepared to move his army closer to Richmond and directed Butler to send 20,000 troops northeast of Richmond to link up with his army. Among them was the 117th New York. The movement began on May 28, 1864, with the regiment marching to City Point where it boarded the steamer* Detroit *for a trip up the York and Pamunkey Rivers to White House Landing. From there they marched to Bottom's Bridge on the Chickahominy River, heading toward Grant's army.*

*Fortunately for Fairhead and his comrades, they arrived too late to be caught up in the deadly struggle at Cold Harbor that would become one of the most lopsided and costly Northern defeats of the war. Reaching Grant after the slaughter of June 2, the regiment manned rifle pits west of Old Cold Harbor. In this position they skirmished for several days, enduring a blazing hot sun and more deadly Confederate artillery fire. By June 10 the regiment had been in the front line at Cold Harbor for a week during which, according to Edward Longacre, the historian of the Army of the James, they "had to endure dangers and discomforts in addition to near-constant sharpshooter and artillery fire. Their rations were so poor and available water so brackish that many, including [Gen.] Baldy Smith, got sick. For more than a week they roasted under the glowing sun. Held on the firing line for hour after hour, they had no opportunity to bathe or change clothing."*[1]

*Frustrated in his attempts to force Lee out into the open where Federal numbers would give them an advantage, Grant decided on another flanking march. This time he intended to shift his army further south as quickly as possible, cross the James River, and attempt to capture Petersburg before Lee could react. Among the first troops to move was Gen. Adelbert Ames' division, including the 117th New York, that once again took ship for Bermuda Hundred. From there, the regiment marched toward Petersburg where it participated in the initial assault on Petersburg Heights.*

---

[1] Edward G. Longacre, *Army of Amateurs: General Benjamin F. Butler and the Army of the James, 1863-1865* (Mechanicsburg, PA: Stackpole Books, 1997), 139.

City Point on the James River south of Richmond was a major Union supply base for the siege of Richmond and Peterburg. *Library of Congress.*

May 29th Sunday. In motion, Va. Made bed. The drizzle had ceased, but the air continued damp. Sky cloudy. Slept very well. Called up at 4 a.m. "Fell in" & moved forward. When near City Point, we halted, stacked arms & made coffee. Then moved forward to the Point, overlooking the landing. The 117th then waited till the remainder of column embarked. 117th embarked at 12 M [noon]. Moved out into the stream and anchored. City Point has been a place of considerable interest and importance. Quite a village and must have been very Picturesque, situated on a prominent Bluff on the right bank of the James River. Made my bed with Corpl Garlic,[1] on a wood pile. Had a good sleep during the p.m. The whole scenery along the James is most beautiful now that nature has clothed, completely, the Earth with fresh verdure. Sad that war should spread its blight over so many spots. Secession has already paid dearly for its work of Originating this conflict. The evil genius of Traitorism must be throttled & crushed, what ever the Price to either

[1] Theodore J. Garlic, a teacher from Vienna, was in Fairhead's Company D.

side. Fiendish hate must be counteracted by Loyal determination. Order to strike tents at 6 p.m. Drizzling rain. We soon had out traps packed & backed with out rubber blankets thrown over our shoulders, covering our knapsacks. Marched off slowly, much halting until the whole column got nicely under way. Moved to Point of Rocks. At 9:00 p.m. crossed Appomatox on Pontoon. Moved onward till 12:30 a.m. when the column halted for rest. Formed column by Brigades & stacked arms. All made beds on the ground, and were soon sleeping. We were tired. The air was very damp, the rain had ceased.

May 30th Monday. At 6 p.m. the Boat, Detroit, moved from its anchorage & at dark the men had laid down to sleep. The decks was completely covered with sleeping Soldiers. One could not set his foot down without touching a Sleeper. The boat cast anchor Early in the night, laid till day break, & then moved onward. Passed the Fortress (Monroe) at 11 a.m. York town at 4 p.m. Kept on our way. Got a mile or two above West Point by dark. Cast anchor. Laid till Day break.

May 31st Tuesday. On board Detroit, York River, Va. Got under way. Many boats in sight. Every boat we pass our boys will Call out "Hardtack." Our rations begin to grow short. The Detroit ran aground and was detained a short time in swinging about to get free again. The Schooner in tow, landed with the horses of the Brigade just rubbed snugly against the starboard side of the stern. The horses were frightened. The hawser was cut, & the Steerman of the Schooner was on his taps to do his part to prevent a [???]. Not more than 20 minutes delay & we were under way again. Schooner in tow. The Banks of the River, and all the country to be seen are dressed in the richest verdure. This section is more uniform than that along the James. Still there is considerable similarity. The Detroit ran aground the 2nd time. We landed at the White House at 11 a.m. Rations were again issued. The whole force was moved to the front or near there expecting our Brig. Col. Alford is "played out" completely. He is a consumate coward. He was ordered to His regt But has succeeded in getting a furlough of six days to Fortress Monroe & will no doubt get a hole of some kind in which to stick the head of a craven.[1] 117th, 142nd & 3rd N.Y. have been united under

---

[1] Here there is an entry in the margin apparently made later — "Col. Alford sent in his resignation. It went as far as Abe Lincoln, was returned from that sound man with his signature as follows. "Dishonorably discharged for cowardice. A. Lincoln." Bully for A. Lincoln."

Brig. Genl. Ames,[1] with 2 other regts. Our regt is glad at the change. Toward evening our Brig moved out to a better location, & put up our tents. 240 men were engaged on fatigue duty. Making the burned R[ail] R[oad] Bridge across the Pamunkey serviceable. The Bridge was completed at Midnight. Our men worked till 9 p.m.

June 1st Wednesday. At White House, Va. Hot with a stirring breeze from the south west. Still in camp (2 p.m.) have lain in camp all day. Wrote to Hannah. Wrote to Mother. Rumors are favorable from all quarters. The men are highly enjoying the bathing in the Pamunkey river. A Portion of the Potomac Army wagon Train came in from the north of the Pamunkey. They are after Forage. Wounded came in them. Army supplies of all kinds are coming on boats to this place. The river for a mile is literally crow[d]ed with sailing crafts of all grades from a canal scow to the largest river steamers. Went to bed at nine o'clock p.m. 5 Co[mpanie]s including "D" were called up at 10 p.m. and went to work throwing up a rifle Pit parallel with our camp. The other regts were engaged the same. In two hours & a half a good rifle pit was up. This was for a protection against Reb cavalry which were in a tight place, & it was not known but they would dash through here. Sergt Rowell[2] and I were working actively & in close proximity. As he brought his spade back & I threw mine forward, the third finger of his left hand was caught, breaking it, without crushing between the 1st and second joint.

June 2nd Thursday. Camp at White House, Va. Regt has been laying at ease to day, washing clothes, writing, washing bodies & shaving. Boats steaming about in the river. Has more of a crowd here than you often see in York Bay. More Wagon Trains are coming in. Some have gone out, going Bottom Bridge way. Droves of Cattle have also gone out. Cattle, horses & mules were unloaded – to day. The Cattle are droped into the river, & then swim ashore. To each horse or mule a small rope is fastened to the Head stall. Then each individual is literally shoved off into the river. They would plunge in out of sight, feet first, rump first, head 1st or all together. In coming up they would flounder, & snort & blow. The man on shore would pull on the rope & thus they would be fished in. There was considerable sport in this, especially the Mules. I

---

1 Adelbert Ames, a May 1861 graduate of the U.S. Military Academy from Maine.

2 Andrew T. Rowell, a mechanic from Sangerfield. He was killed on October 27, 1864, when a shell fragment broke his skull. Fairhead considered him "a bright, good soldier." Veterans in Waterville named Rowell Post, No. 23, Grand Army of the Republic, in his honor on May 23, 1868.

stopped a few moments and looked at this manner of unloading. Wharfs are scarce, but are being extemporizes with magic rapidity. Everything about the landing has the air & bustle of an extensive mart of commerce. Some Rebs were brought in to day. Female & Male "Contrabands"[1] came in, in droves with the Trains. They universally have a large bundle on their heads. They are a dirty, ragged set. I heard one of these women swear like a traitor. Upward of 600 Prisoners were brought in to day. They are a motley crew. Rainy at evening.

Lt. Isaac Dann, Company A. *US Army Heritage Education Center.*

June 3rd Friday. Drizzling rain during the morning ceased at 8 a.m. Some sultry during the remainder of the day. Broke camp at 10 a.m. Moved away from White House at 1 p.m. Heavily loaded with rations. Pulled steadily for the front. Met large numbers of slightly wounded walking to the rear. The badly wounded were being carried in the wagon Train, going to white House for supplies. Marched till after 11 p.m. laid down by road side to get a little rest. Could hear the popping of the Pickets as they occasionally fired.

June 4th Saturday. Cold Harbor, Va. Waked up at 2 a.m. Got under way, & before daylight had taken position in a forest to the rear of our rifle Pits. Sergt Evans[2] was used up by the March Yesterday. I took charge of the Co[mpany] in his place. Two of the boys were also worn out. We marched near 20 miles. During the day a detail from our regt commenced rifle pits. At evening the regt moved into them. The place in the forest was considerable dangerous. A few men were wounded. Lieut Dann[3] was wounded through the right breast. During the night we worked more on the pits. The Rebs made a demonstration in our front, were repulsed easily. Our line of troops did not open on the Rebs. The 1st line being sufficient. Our cannon sent in a number of shell most savagely. Rain during night & day.

---

[1] Escaped slaves.

[2] William J. Evans, a textile mill spinner from Utica.

[3] Lt. Isaac H. Dann of Clinton died the same day.

June 5th Sunday. In rifle pits, Cold Harbor, Va. Cloudy with some rain at early morn. The Rebs opened on our Batteries which answered briskly. Clouds broke & sun shone warmly at 3 p.m. The art[iller]y did not last but a little while. The Reb sharp shooters have been active to day. But all our troops have thrown up Earth Protection. Our men are safer than yesterday. A Detail of nearly 100 good shots from Brigade were sent out to the front pits to work against the Reb S. S. [sharpshooters] who are less active than before this detail were sent out. Report says Lt. Dann is dead. A true report. Sharp Shooting was active during p.m. 1 of our (142nd regt) men was killed. 3 wounded. Some of the others got pecked. One in "K" Co. received a severe wound in the chest, passing from the shoulder forward to the pit of the stomach & lodging. The constant cracking of S. S. Muskets makes a disagreeable rattle. The Arty once in a while would send out a few shells. A Reb Bat[tery] which has a correct range on the section sent a number of shells very close to us, which burst, cutting limbs twigs & trees. No damage was done to our regt. Some from other regts were injured. Only a few. Sun went down nicely. Evening clear. All s. shooting had stopped. The quiet was profound, & contrasted greatly with the rattle of the day. An onerous stillness. At 8 a Musket shot. Two or three. a roar. vollies. The rebs charged on the 2nd Corps left of us. The charge lasted some time. Rebs repulsed. Our Arty opened vigorously. The Rebs yelled some as they continued their work. Our Boys cheered some. The Storm subsided. We sat with Musket in hand ready for any thing, not Knowing but that the charge would break out afresh, as the pickets were pecking away a little. There was also a show of charge to the right of us. A Brass band among the rebs to their rear set up their music as the charge ended. They continued some time. This was listened to with ominous silence. All became quiet & our boys laid down & went to sleep as best they could, guns near at hand. Some little shooting along the lines. Joseph Hepworth, White, Roberts and Clowenger of the 14th Heavy Arty came to see us today.

June 6th Monday. In rifle pits, Cold Harbor, Va. Sun shine & breeze. Still in these Rifle Pits. Put up our shelter tents like awnings to keep off the rays of the sun. S[harp] S[hooters] commenced at day light. Our Boys in the front rifle pits are popping away. Some Arty shreaks away on its death's mission but they oftener miss than hit, the same as it is with bullets. John Beaver[1] came to see the boys. He looks lank but

---

[1] This may have been the father, or a relative, of Edward Beaver, a farmer from Rome wounded at Drewry's Bluff who later died of his wound at Bermuda Hundred.

Benjamin F. Miller, Company I.
*US Army Heritage Education Center.*

tough. The whole of us are some worn, dirty but resolute. We will stick to it till we win back the honor of the Flag. We are hardy. At times we get short of rations on acct of so many rapid & sudden moves much is left behind. Wrote to Minnie while in the pits. S[harp] S[hooters] continued in the Pits till after dark, were then relieved, & moved back into the woods. Laid down having formed "Columns by Div[ision] right in front."

June 7th Tuesday. Camped in woods, Cold Harbor, Va. Pleasant. Was waked at early day light by the crack of a shell. A number followed, going completely over the forest in which we were stationed. We got into a more regular column, cleared up the ground a little, & then put up tents. I got a good sleep. Mail came (large). I got 2 letters from Minnie, 1 letter from Ed Hughes, 2 from Mother, one from Annie. A paper box also came from Mother, 5 sticks liquorice, 5 bottles of diarrhea drops, 3 cakes, maple sugar, 1 Pack Envelopes, court Plaster, Pills. Went to the 10 NY Arty. Saw W. E. Ackroyd, Charley Overocker, Harris. Came back to camp. The Rebs sent in a number of shells, a few bursted near. Lt. Miller[1] who was detailed as assistant Provost Marshal & who was near that section, got his leg broke from one of the shell or splinter. Toward evening we got orders to break camp & get ready for the Pits again. At dark moved to them. Went into the front line of pits. Sent Pickets or Videtts in front of pits. They laid forward some distance, flat on their fronts. We in the Pits slept but little. A few shots from scattering rebs. Cool all night, more than usual. No demonstrations during night.

June 8th Wednesday. In front line of Pits, Cold Harbor, Va. Pleasant, some cloudy, counteracting the direct effects of the sun's rays. Men make themselves jovial & loung[e] away the time, ready for a moments call. Some S. S. Our men are ordered not to shoot. Tis a useless practice

[1] Benjamin F. Miller, a Utica farmer, was wounded on June 4, 1864, and discharged for wounds on December 17.

as nearly Every shot is made at random or in the direction of some person supposed to be in the locality arrived at. One or two or sometimes more become injured by this practice, but this does not materially weaken either army. Servey[1] in "E" was shot through the neck this forenoon, Killed instantly. He was reckless in exposing himself above the Pits. We were relieved after dark. Came to this section where we laid night before last. Laid down & went to sleep. Had very good bed. Star light. Could see them by looking through the openings among the trees.

June 9th Thursday, Camp in woods, Cold Harbor, Va. Pleasant with a stirring breeze. Woke at daylight. Sleep was refreshing. The best nights rest we have had in more than a week. Inspection at 9 a.m. John came back today.[2] Wrote to wife. Wrote to Sisters Annie & Mary jointly. Saw Theo. Ashley, of the 24th Cavalry, Hospl Steward, went, also, to the 14th H. Arty. Saw J. Hepworth & Clomenger.

June 10th Friday. Camped in woods, Cold Harbor, Va. Fair & breezy. Were allowed to remain in camp another night, and we did good service in sleeping. Cobb, Hand, McAdam, Rose & Swan came from the 2nd H. Arty to see us. They look well. Swan has a grazed sore on the side of his forehead. The result of a friendly ball. Went into the Pits after dark (near line). Half of the men were allowed to sleep at a time. I was up the first half of the night. Quiet.

June 11th Saturday. In the Pits, Cold Harbor, Va. Warm. Put up our shelters for awnings across the pits to exclude suns rays, read some papers. Mail came bringing one letter & three papers. 2nd Mail brought a Dollar Monthly. Read testament. Sharpshooting going on. None in 117th injured. Rebs sent over a few mortar shells during the after part of the day. Our Batteries replied with some vigor. Uncle came to the Regt to day. Had a long conversation with him. His time is out in 19 days. Goes home. No relief came in, we arranged for remaining another night. Half awake at a time. Some little firing during the early night. All quiet there after.

June 12th Sunday. Trenches, Cold Harbor, Va. Broken, dark clouds; warm sunshine, fresh breeze. S. S. going on. Men in the Pits are busy writing letters, reading letters, papers, magazines &c. Some are sleep-

---

[1] William H. Servey was a bartender from Rome.

[2] John was his brother.

ing, some playing cards. Some demurely, or closely thinking. Some who have few abilities to use their time advantageously, are trying to be jocular, but are so insipid & pointless that their blabadash has such an effect as to often produce mental nausea. We get our meals of hard tack, meat & coffee, & set down to use the time as best we can. Wearily drags the time after. Orders to move kept us on the guard from 10 a.m. till 10 p.m. when we left the Pits. All the force had fallen back. At 10:30 p.m. we formed column & moved forward or rather to the rear. Passed the 6th Corps behind Earthworks & tramped onward, arriving at the White House at 6 a.m.

June 13th Monday. en rout[e] for Bermuda, Va. Roads excessively dusty. Could feel the dust in the nasal passages as one breathed. Halted to rest but very little. Men greatly jaded when we got to White House. Took boat U. S. M. at 9 a.m. (Had time to wash body in the Pamunkey.) Moved from wharf at 9 a.m. Made West Point at 1 p.m. Passed Yorktown at 4 p.m. As soon as the boat made off the men laid down, completely covering the decks, and went to sleep. Passed into the Bay at 6:30 p.m. Moved to New Port News at about 12 mid night.

June 14th. Pleasant. At 3 p.m. Debarked at Bermuda Hundred. Moved to an old Camp in which we remained the night before going on the trip to White House. Put up tents. Went to bed immediately after "Tatoo."

June 15th Wednesday. Hot & dusty. Roused at 2 a.m. Orders to strike tents & be ready to move. Got ready. Made coffee, laid down, knapsack for pillow. Started at early day break. 18th Corps mostly under motion. Crossed the Appomattox on the Pontoon bridge, & forward to the rear of Petersburgh Va. where we formed battle line about noon. "D" & "H" were sent out as skirmishers. Deployed under cover of the wood then pressed forward wheeling to right considerably. Throwing us out of the woods into an open field & a portion of slashing. We were very close to the Enemies lines & chain of Defences. They had much Arty & used it vigorously all afternoon. The shot & shell with grape flew thick. We could distinctly hear the Arty commanders give their orders to load & fire. Harry Balis[1] was Killed on the skirmish line, shot through the head about 6 o.clock p.m. Soon after we moved forward, the whole Picket line on a charge for the reb works. Short, Hot & successful. "D" & "H" snatched 150 graybacks & 5 pieces of Arty. The Negroe troops on our left charge forward with much impetuosity. The whole line of outer

[1] William H. H. "Harry" Balis was a marble cutter from New York Mills.

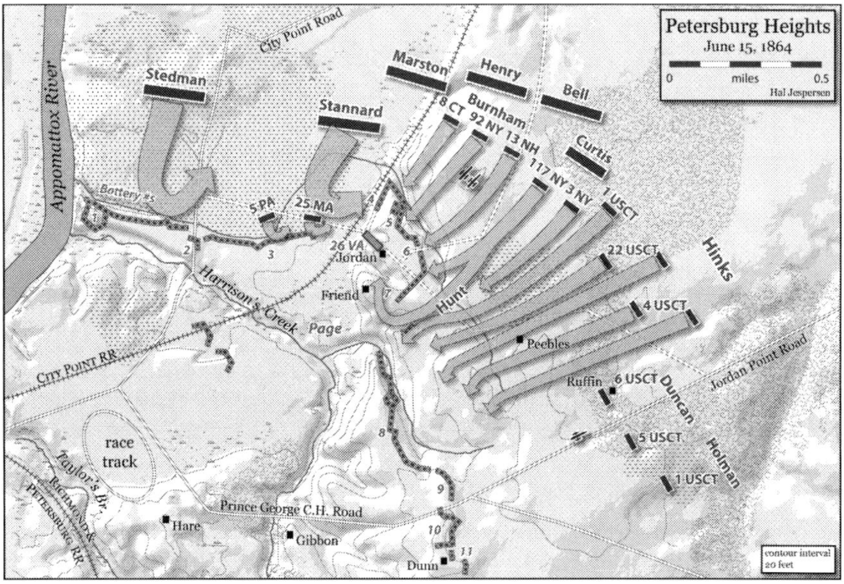


Capt. William J. Hunt led his Company D including Fairhead and Company H under Capt. Almon R. Stevens, about 100 men, as skirmishers in front of the attack. His small force was able to slip through a ravine and fire into Batteries 5 and 6 from behind causing them to surrender and opening the Confederate defenses to a successful assault. *Hal Jespersen.*

---

Defences along our front was carried. 17 Pieces of artillery in all were captured & 250 prisoners. The 117th with other regts. came forward immediately as support. Gen. Smith was down & into the Defences very soon after we had taken them. Three of the pieces taken were reversed[1] & discharged at the retreating rebs. Our arty came into play as we started on the charged & served their hail in an active Manner over our heads. 4 of "D" Co. were wounded in this charge. I prevented a Negro soldier from Killing a prisoner: retaliation for Fort Pillow.[2] Rested in the woods during the night.

---

[1] The artillery was turned around to use against its former owners.

[2] On April 12, 1864, African-American troops attempting to surrender at Fort Pillow, Tennessee, including those wounded, were massacred by Confederates under Gen. Nathan Bedford Forrest.

Capt. William J. Hunt, Company D, on left, and Capt. Almon R. Stevens, Company H, on right. *Donald Wisnoski.*

---

June 16th Thursday. In Defences of Petersburg, Va. Hot with wind. The Regt. Laid near Battery No. 9, Petersburg defences. At 5 p.m. Troops packed up close ready for move or action. Troops moved forward, taking new stations nearer the Rebs. Our Arty opened in a most furious manner, causing clouds of smoke about us, & clouds of dust & smoke where the shell & shot stuck & exploded. The sound was an incessant Explosive roar. Just at dark the 117th & brigade moved off going to the left somewhat and forms a second Battle line behind a line of the 2nd Corps which forced the rebs back to an intermediate line. Here we laid on our fire locks all night. Were roused three or four times by a spasmodic rattle of musketry. The front line threw up a rifle pit during the night.

June 17th Friday. In defences of Petersburg Va. Hot with breeze. Still in battle line. At early light we commenced to throw up a rifle Pit. In 3 hours we had a roomy & heavy pit completed & [???] strewed over the bottom. Remained about an hour. Ordered to move out, a Brigade from 2nd Corps taking our place. Our Brig 3rd moved to Reb Batt No. 10 taking position near it with "D" & "H" <u>on</u> the Battery itself, a very commanding place. We have reversed the works enough to accommodate us for good Pits. A Six gun Batt Brass is in position beside the old work. In front of us is a low dale wholly cultivated, mostly planted with corn. Both Rebs & Feds have pits on this field. Heavy canonading on our left at 5 p.m. Canon near us speak once in a while. Capt. Stone[1] "B" Co. was killed while walking along the top of a rifle Pit, running parallel & from the side of the Earthwork. The ball Entered at his ear. A reb S. Shooter. This being so high a position, the Rebs can throw bullets here without much exposure to themselves. Wrote to wife yesterday. Wrote to Harry Balis' Father yesterday concerning his [Henry's] death. At 9 p.m. we moved to the ret & started back to Bermuda 100s.

Capt. Josiah Parsons Stone, Company B. *US Army Heritage Education Center.*

---

*The performance of the 117th New York at Petersburg Heights was every bit as impressive as it had been at Drewry's Bluff. Under heavy fire, it reached the Confederate fortifications where Capt. William Hunt found the small ravine leading to the rear of the defenses. The rest of the regiment followed, its colors the first to be planted on that portion of the Rebel fortifications. When a final accounting occurred, the regiment had captured a four-gun battery along with the colors of the 26th Virginia and 211 prisoners including the unit's commander, Lt. Col. James C. Councill, two majors, four captains, and nine lieutenants. The regiment's breakthrough completely unhinged the Confederate defensive line leading to its collapse. The assault cost the regiment an estimated 54 killed, wounded, and missing.*[2]

---

[1] Josiah Parsons Stone was a Camden attorney. Veterans in his hometown named J. Parsons Stone Post 482, GAR, in his honor.

[2] Howard Thomas, *Boys in Blue from the Adirondack Foothills* (Prospect, NY: Prospect Books, 1960), 208; James A. Mowris, *A History of the One hundred and Seventeenth Regiment N.Y. Volunteers (Fourth Oneida), from the Date of its Organization, August, 1862, Till that of its Muster Out* (Hartford: Case, Lockwood & Co., 1866), 113.

# Petersburg

## June 18-July 31, 1864

*Before Grant could bring the full weight of his army to bear on Richmond and Petersburg, reinforcements arrived from Lee's army and from North Carolina to bolster the defenders. With the defenses too formidable for a direct assault, Grant decided to change strategy, gradually using his numerical advantage to extend his siege lines south and then west of Petersburg to gradually cut off all of the road and rail lines into the city from that direction. This would isolate both Petersburg and Richmond from any reinforcements, provisions, or supplies. As a consequence, the 117th New York, assigned to the 1st Brigade, 2nd Division, Tenth Corps, moved first into the northern extremity of the Bermuda Hundred fortifications with the 112th New York directly next to the James River, the 117th New York being in support. The day after its arrival, Fairhead's regiment moved into the first line of entrenchments where it remained until June 27 when its brigade moved into the siege lines around Petersburg. Positioned close to the Appomattox River, the men were withing easy sight of the city whose church bells they could clearly hear.*

*For the next six weeks, the regiment rotated in and out of the front line, exposed almost daily to the summer heat and humidity, inadequate food, the attacks of vicious vermin, deadly sharpshooters, exploding shells, and occasional attempts to penetrate their lines. Johnny Foote in Company C explained that it was "one of the most uncomfortable places on earth. Nothing to protect us from the burning sun but our shelter tents, it seemed as if we should surely melt & as sure as a man exposed himself in the least, just so sure would a bullet whiz after him so that to draw rations or get water [he] many have been shot while doing such things; and besides we had scarcely no sleep at all for in the day time it was too hot, & at night, Bombshells were flying to[o] near, or else there was some scare so that we didn't get any then. Such is life in the trenches."*[1]

---

June 18th. Reach Bermuda 100 at 3 a.m. Got into column & went to Sleep. Waked at 9 a.m. The engagement which raged all night at the front continued till near 2 O'clock p.m. Terrific cannonading & musketry. We laid in camp all day. Made a package of Harry Balis' Effects &

---

[1] Letter, Foote to sister, July 6, 1864, Foote Correspondence, Duke University.

sent them by Express to Utica.

June 19th Sunday. Bermuda Hundred, Va. So quiet that it seemed peculiar. Had Co. Inspection. Wrote to Mr. Balis concerning the Package. At 3 p.m. moved to the fortifications, & put up tents just behind them.

June 20th Monday. Bermuda Hundred, Va. Excellent weather. Got letter from home one from wife. Wrote to Minnie. Perfect quiet. Stewart Walcott[1] came to see the men from our Place. He is associated with the Sanitary Commission[2] department in the 18th & 10th Corps. He gave John[3] & me a lemon each.

June 21st Tuesday. Fortifications Bermuda 100, Va. Foggy during morning. Hot all day. Our Monitors[4] & Reb rams & one Baty have exchanged a number of shots to day. We have laid quiet. Wrote for S. Sharp. Washed & mended shirt. Some exchange of shots between our Gun boats on the James, & the Rebs. Later Part of p.m. our regt moved & took position at Battery No. 5.

June 22nd Wednesday. Clear & bright. Coolish & cloudy at morn. Hot at p.m. Our Boys have fitted up in good order. Boards were here that were used by those who left. President Lincoln passed our camp about noon. Dark suit. Plug hat. Rode a large bay horse. Our regt cheered heartily Old Honest Abe all full of enthusiasm. Received the act by taking off his hat. We were glad to see him. Men all eager to get a good view. Wrote to wife. Wrote to Father. Detailed to Picket at 7 p.m. 150 men from Regt. Our Position was on the outer edge of the slashing in front of fortifications. Had a snarley job in finding posts.

June 23rd Wednesday [Actually Thursday]. On Picket in front of Fortifications. Hot Sun, a slight breeze. Nothing unusual through last night. We are not in close proximity to the Rebs. Can hear them halloo to each other once in a while. About 5 p.m. the Rebs opened their Batteries. Ours remained quiet for a while, & then opening, soon dried up the

---

1 Walcott was a member of the Walcott family that owned the textile factories in New York Mills.

2 The United States Sanitary Commission was a relief organization created in June 1861 to support sick and wounded soldiers.

3 Fairhead's brother, John S. Fairhead, a dresser in Walcott & Campbell in New York Mills, was a drummer and the regiment's principal musician.

4 A steam-powered ironclad warship.

Alfred R. Waud sketched this picket post with Confederate fortifications in the background. Note the covering to provide shade from the sun. *Library of Congress.*

Johnnies. We sent in 3 or 4 last shots. 2 of "C" Co. Boys who went forward from the line to exchange coffee for tobacco with the Gray backs were not fortunate, & were no doubt taken Prisoners. Expected to be relieved at 6 p.m. Relief did not come.

June 24th Friday. On Picket in front of Fortifications. Sun hot, with breeze that at time felt as though it were a blast from a furnace. This morn hear that the Troops have gone from the fortifications. Every man that can possibly [be] spared. 117th gone. Picket all day. Quiet in front of us. Some shots exchange on the right.

June 25th Saturday. On Picket in front of Fortifications. A blazing, parching sun. Still on Picket. 5 Johnnies came into day. One man from the 142nd deserted to the Rebs just at eve. We were ordered to be doubly vigilant, fearing that the deserter might give information of our force.

June 26th Sunday. On Picket in front of Fortifications. A little Breeze stirring in the Morning, which from the red hot rays of the sun, bids fair to become almost stiffling by noon. All quiet during night. I believe that the Johnnies are not very numerous on our front. They sing & halloo considerably during the Evenings, but it sounds more like being done for the purpose of deceiving than as a mere outburst of jubilant spirits. We hear cannonading from the direction of Petersburg each day. Some times quite active, but not of a character as to lead us to think there is a general engagement. Frequently hear the detonating boom this morn. The 6th & 7th Conn Regts came back to the Fortifications last night. Blazing hot all day, till near 5 p.m. a Thunder shower, not much rain fell in this section. No rain during night.

June 27th Monday. On Picket in front of Fortifications. Sun rose clear. The little rain yesterday made the air of this morn some what refreshing, but it became hot by noon. Soon after the dark, foreboding thunder heads came rolling up from the N. W. During the last few days the sun has felt in its radiations much like an immense molten mass of metal. At 12 midnight last we were all ordered to be under arms. Rebs reported to have rec[eive]d large reinforcements in our front. All continued quiet. We have slept considerable (by turns) during the day. Heavy thunder at later part of p.m. Much wind, but little rain. Move the line (in one section) some what, so as to make closer posts. Quiet all night.

June 28th Tuesday. On Picket in front of Fortifications. Cool air last night, & this morn. The men on duty fixed their Posts by throwing up a slight earthwork. Finished them about 9 a.m. A little after we heard of a relief, & that we were to go to the regt. Relieved at noon, went into fortifications, got dinner. Rested till 4 p.m. Started for Petersburg, excessively dusty. Reached Peters'g at 8:30 p.m. Did not go to the Regt. Laid behind a captured reb rifle pit.

June 29th Wednesday. Near Petersburg, Va. Hot, with a freshest breeze. Cannon boom & rifle crack woke us at times during the night, still we had a comparative good nights rest. Up at sun rise. Wrote to Hannah. Got letter from. Went to regt at noon. Lt. Denton[1] com'd'g Co. Capt Hunt[2] wounded yesterday in knee. Sharp Shooting. Good rifle pits. Some boys in the regt have been wounded since being in this Pit. They are careless in exposing themselves.

June 30th. Thursday. Pits before Petersburg, Va. Hot. Comparatively

---

1 Alonzo Denton, a clerk from Remsen suffered a serious wound and was later discharged for disability.

2 William J. Hunt from Trenton died from his wound on July 31, 1864. Veterans named William J. Hunt Post 510, GAR, in Holland Patent in his honor.

quiet during the night. Worked at the muster rolls some. Some offensive movements, in which the men did not become much exposed to the Reb fire. The Rebs stirred about some. We could see the dust they kicked up. Our regt moved into the Picket rifle pits, keeping close. Before dark we moved back to our Station in Pits. Co. Mustered about 2 p.m.

2nd Lt. Alonzo Denton, Company K.
*Oneida County History Center.*

July 1st Friday. Pits before Petersburg, Va. Very hot. Some musketry, cracking of arty during night. Had some little sleep Half Co. being on the look out at a time. Exchanged Pits with 112th N.Y. they coming into front Pits we moving to the back line. Posted one man for alarm Guard. About 11 p.m. active musketry roused us. Each man put on his accoutrements & took his gun. No advance of force on either side. Our Pickets & a fatigue party advanced & started & possessed a new pit some little in advance. Quieted down in about an hour.

July 2nd Saturday. Rear Pits before Petersburg, Va. Hot as usual. Washed Pants, Shirt & Kerchief. Also washed body. Beligerent affairs as heretofore. Got letter from wife answered the same. Sergt Rowell came back to regt on the 28th ult. having been away 28 days. His finger is healed nicely. Pvt. Berkely also came. He has been away from us about 8 months sick.

July 3rd Sunday. Rear Pits before Petersburg. Hot. Some clouds at intervals. Very quiet during night, last. Was selected as commissary sergt for co. to day. Our Boys like the arrangement. Chaplain of the 112th N.Y. gave out a short sermon at 6 p.m. Text Proverbs 14-26th "The fear of the Lord" &c. Stewart Walcott came to the company, bringing Farina, corn starch, condensed milk, lemons, oranges & some other small articles. Our boys all made their black tin pails full of rice pudding. They are benefitted in stomach & spirit by these treats. They came from

This Alfred R. Waud sketch shows sharpshooters positioned behind fortifications. Note how the embankment is high enough to protect the head of the man on the right, with a small open box allowing him to fire with minimal exposure. *Library of Congress.*

the Sanitary Commission. S. Walcott shows an earnest active kindness toward our co. The men from the mills being the influencing force that draws his regard. Went into the front pits to night, relieving the 112th N.Y. Considerable quiet. A number of Whitworth[1] solid shot were shot at our lines at sundown, some what peculiar to throw such at lines, & that at random. Did little or no injury near us.

July 4th Monday. In front Pits before Petersburg, Va. Most of day the sun was obscured by clouds, fresh breezes. Many of our men expected there would be a large amount of Arty firing during day. But little, at times, a little active. Some men in the regt were hit yesterday. 1 Killed.

[1] The Whitworth artillery piece was a rifled gun made in England. It was one of the first to be loaded at the breach and was exceptionally accurate.

Taken along the Petersburg trenches, this photograph illustrated how the troops stretched their shelter tends and blankets across the ditches for protection from the sun and rain. *Library of Congress.*

---

6 wounded. One of our Mortar shells dropped in the Reb rifle pit. One Reb was thrown 15 feet above & over the Parapet. His clothes were on fire. Dead. The Rebs could not get him till after dark, as should any attempt be made our sharp shooters would pick off the venturesome men. Some time since 5 of the Union soldiers volunteered to remove some of our wounded from the front. Each of them was shot dead. The horrid realities of war are barbarous in the extreme. Gracious Creator give Early success to the Right and discomfort to all that is wrong. H. W. Garrett of S. Trenton came to our co. to see me. I was exceedingly glad to have a counter privilege. He is hale & spirited.

July 5th Tuesday. In front Pits before Petersburg, Va. Sun clear, stirring breeze. Some Picket firing during night. Rebs throw some arty. Our arty usually answer. Stillness has been disturbed about as usual. Some casualties. ½ gill of whiskey drawn each day for each man <u>in</u> the pits. Were relieved by 112th N.Y. at eve. Moved some to near the Brig Hospl

Stand. Laid down & went to rest.

July 6th Wednesday. Petersburg, Va. Some airy with clouds, but warm. Wrote to wife. While we lay back to day, the Rebs sent in a number of Whitworth solid shot. They flew with their terriffic screaming rush rather promiscuously along & near the road. One man in the 3rd N.Y. was Killed. This was all the injuries from the Shot. Drew new clothing. At evening our regt again entered the Pits relieving the 112th N.Y. About 10 p.m. the Rebs sent over as many as 40 Shell. Our Batts remained silent a long time. The Rebs seemed to be striving to draw out our fire. They seemed to become satisfied & quit firing. Then our Batts shelled the Reb lines vigorously for half an hour. This was near our immediate front. Very quiet the balance of the night. A slight random Picket firing to the left.

July 7th Thursday. Petersburg, Va. Stirring breeze from the west. Sun clear. Considerable rifle popping this morn. Some arty Dueling during day. A shell struck and bursted in the pit occupied by "I" co. wounding one, severely, wounding one, badly wounding one, & slightly wounding one. Total 4. One in "B" Co. kill[ed] by rim of our own shell. Reported dead. One in "F" co. got a hard contusion on shoulder. Amelious Ackroyd came to see me to day.

July 8th Friday. Petersburg, Va. Hot & Sultry, clouds lazily floating through the sky. Thunder & a little rain yesterday. Nothing unusual to day thus far, 3 p.m. Got letter from Mother, also, one from wife. Wrote to Mother. In trenches. The Rebs made a slight demonstration toward 5 p.m. Our Batts & Pickets were active in a moment. Our men in the Trenches were immediately on their toes. The Rebs hid themselves. All became quiet again. Left Trenches at 8 p.m.

July 9th Saturday. Petersburg, Va. Hot & some wind. Had good nights rest, laid in the wood where we located when we were out of the Trenches. Got letter form Wife answered the same.

July 10th Sunday. Petersburg, Va. Hot & windy, cooler night. Drew two days rations. Was busy nearly all day with drawing, & issuing the rations to the men. We are having plenty, and good rations. This work brings many peculiar Sabbath duties. How blessed will it be for Earth "When Nations shall learn war no more." Went into the front Pits after dark. Negro Troops lay in the Second line of Pits just in rear of our Brigade. They hold them steadily, not doing duty in the front Pits near us if anywhere.

July 11th Monday. Petersburg, Va. Hot, Sultry but with some stirring air. Thunder heads float lazily around the horizons. Some distant (to the S. W.) reverberations greeted our ears during the afternoon. The Earth is dry, Parched and excessively dust. A thorough soaking would do us good. Our Batts have vigorously shelled the Rebs to day, but have not recd much reply. Toward 4 p.m. the Johnnies throw over a few shell. The news of the Engagement Between the Kearsarge and Alabama is received with much gratification by the men. The Dastardly Cowardly Merchantman Robber, has met with a wild fate at the hands of the 1st war vessel she has had courage to fight with. She was forced to [???]. Her dash with the Hatterass over one year ago was a piece of arrant Treachery in the Alabama which latter hung out false colors. May the fishes of the waters where she sunk, not be poisoned by the perfidious cowardice that has marked her career.[1] Heavenly Parent, Do thou Grant Order that all Evil Schemes or developed projects now among the People Either North or South, Shall sink beneath the waves of Truth and Righteousness. May thy Will be done on earth as it is in Heaven. Come Sweet angel of Purity and Peace.

July 12th Tuesday. Petersburg, Va. Hot and windy. In the Trenches. Some sharp shooting also arty during day as usual. We go to the rear after dark, had a good night's rest. News of the invasion Northward of the Rebs makes speculation for our boys. The Raid is causing some fright among non combattants, women and children.

July 13th Wednesday. Petersburg, Va. Hot and Sultry. Boys busy washing clothes and body. At night went into 2nd line of pits, laid down for rest, one man being on duty. The Rebs opened one of their battery very vigorously, ours opened and had a sharp duel & the Rebs quit 1st.

July 14th Thursday. Petersburg, Va. Hot. Laid in the 2nd pit all day. Went into 1st after dusk. Wrote to wife. The Rebs tried their battery again. Our battery set in as usual. The Rebs dried up 1st.

July 15th Friday. Petersburg, Va. Hot, but since the new moon the nights have been cool & makes them better for rest. The Rebs on the right opened upon our Batts. The result was a terrific Arty fight in which our men as usual compelled the Johnnies [to] shut up. I was on a hill where I could see the smoke of the discharges. On the Reb side

---

[1] The CSS *Alabama* was a commerce raider whose mission it was to sink Northern merchant shipping. It was caught and sunk off France by the USS *Kearsarge*.

there was far more dust than smoke. It lasted nearly ¾ of an hour. Toward night the Batts near us had a set to. Our Batts beginning & ending it. One of our shells exploded something over among the Johnnies. A dense cloud of smoke rose at the bursting of the shell. Another shell sent a reb a few feet above the Parapet. All quiet through the night. Only a few musket shots were fired through the Evening. A beautiful moon light night.

July 16th Saturday. Petersburg, Va. Coolish morning. Some musket popping. The Rebs have been at work on their parapets during the night. Our gun have sent in a number of saucy shots this morn. Our Pickets & the Rebs exchanged papers this morning. The Rebs are getting some rough usage about Washington & vicinity. They will no doubt be a sorry set by the time the demonstration ends. Came into the 2nd line of pits after darkness had set in. Quiet during night. Slept well. Coolish night.

July 17th Sunday. Near Petersburg, Va. Stirring wind from the N. E. has come from that Quarter for a week, varying a trifle at early morning. Sun hot. Too all things as easy. Very quiet. Only a few discharges during the day. Mail came. 2 letters from home to John. A note for me from Sister Mary. <u>Wrote</u> <u>to</u> <u>wife</u>. At Evening the Brig of Negros came back they going into the 2nd line, we falling back to the pine woods. Laid down for sleep.

July 18th Monday. Near Petersburg, Va. Coolish. Evening cloudy with spots of rain. Went to front pits at dark. Very little Arty work during the day & night. S. Shooting going on to some extent.

July 19th Tuesday. Near Petersburg, Va. Morning cloudy with cool wind, little rain. In Trenches rain began to fall steadily about 8 a.m. The position occupied by "D" & "E" Cos. in the Trenches is low, the water running from the Trenches on the hill settled in "D" & "E" Co. quarters. We remained until after M. when the mud & water became Knee deep, when we moved to the rear, Still holding ourselves in readiness to move to the Trenches in a jiff. The rain Eased up. We had our shelters up. Clouds continued dense. Had good night's rest. A few sprinkles during night.

July 20th Wednesday. Near Petersburg, Va. Waked after daylight. Clouds still heavy but no rain, remained out of Trenches. Regt came out to the rear at dark. Coolish with but little sunshine.

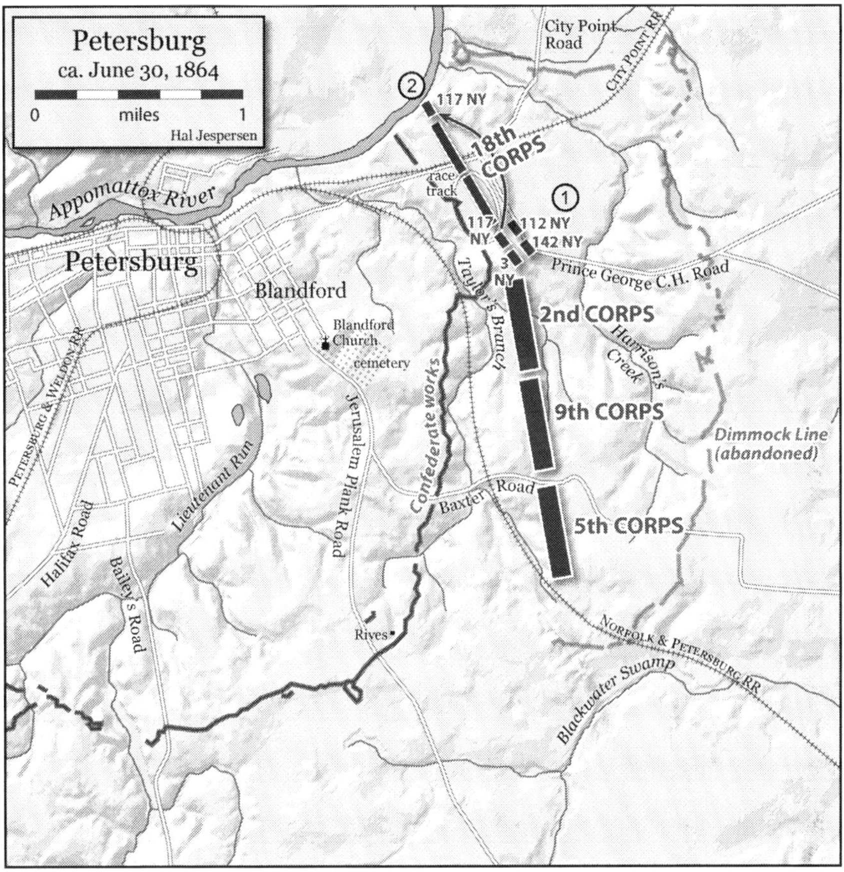


① The position held by Curtis's brigade on June 30, 1864. ② The later position of the 117th New York next to the Appomattox River. *Hal Jespersen.*

---

July 21st Thursday. Pleasantly cool, some clouds. Quiet as usual. In the rear washing & cleaning as usual. Got letter in John's from Mary. <u>Wrote</u> to <u>Mary</u>.

July 22nd Friday. Near Petersburg, Va. A brisk wind & cool. Sun warm but nothing like oppressive. Regt in the rear. At 11 p.m., last night, the rebs opened one of their Batts. Our bats opened a very shattering concentrated fire upon the Johnnies' guns. They sent a few back, but our fire was too hot, & the "Gray" dried up.

July 23rd Saturday. Near Petersburg, Va. Coolish, salubrious breeze. Some clouds which partly shut out the sun's rays. One of our Mortar Batts is statedly throwing its shells over into Secessia. The Rebs do not often answer. They pop away with their rifles, but do very little damage as our men are very chary[1] about exposing themselves. Our Pickets and S[harp] S[hooters] reply when inclined. ½ gill whiskey rations are regularly issued to the men. Got letter from my Minnie. Wrote to wife. In Trenches. Quiet as usual.

July 24th Sunday. Near Petersburg, Va. Continuous clouds cool breeze from the Northward. In Trenches. Drew two days rations. A regt 142nd [N.Y.] had service at 4:30 p.m. Brass band made music playing Old Hundred. A Fatigue detail has been at work cleaning the section we occupy when in the rear. New arrangements having been made which divides our Brig into 3 reliefs, thus allowing us 4 days out & 2 in the Trenches. We expect to drill & pay attention to the more particular Phases of camp military life. Sergt Clarke[2] had 6 of our co. at work on the section to be occupied by "D." The Reb Sharp S. are sending bullets quite low. One man was passing to the rear, a little to the right of our Co. A reb S. S. shot him through the body, entering at the back near the spine, going through him. He dropped down after turning & looking at the reb works. They shoot through camp holes. The man died in about 10 min[utes]. This man was very incautious in going across the field in the place he attempted. A number of men have been wounded in this same locality. Moved out of the Trenches at early dusk; went to the selected position; put up tents hastily; rain commenced falling soon after; dug small trenches about the tent which was on a some what low spot. About midnight was waked by water coming in an intrusive manner too close to my shoulder for comfort. The small trench had filled and the surplus had worked partly under us. Shifted a little to the center of the tent & went to sleep [but] was again waked by the wind flapping my tent. Got up, drove the pins anew. Went to sleep again. Suddenly roused again by the falling of one end of the tent pole. Raining hard & wind strong. Moved "stem" 1st out from under the tent. Berkles & I fixed up the tent. Laid down damp & chilly. Spread our woolen blanket (wet) over us and went to sleep again. Now slept till day light. How pleased to see light after such a thorough experience in Soldier's comfort. Boys were all jocular soakings. Nearly ever one got a little, some much. Rain gradually ceased as morning grew into [day].

---

1 Reluctant.

2 Hermon Clarke was from Waterville.

July 25th Monday. Near Petersburg, Va. Arranging camp began about 8 a.m. By 11 a.m. a regular military village was developed. This the first regularly pitched camp the Regt has formed for 33 days. Dress Parade at 5 p.m. An Order which was read, christened our Camp with the cognomen Curtis in honor of Col. Curtis[1] com'd'g the Brig. Curtis is a man who stand full 6 feet 7 inches in his stockings; he is lean at that and has the appearance of being much taller. J. B. Spictor & Samuel R. Campbell[2] reached here today, one week from New York Mills. It is cheering to get a glance of old friends. They report all well at Home. They came out here in the Service of the Christian Commission. Considerable popping at the front. Some Arty. The Rebs have pitched some shells over near the left of our Div. & right of the 9th A[rmy] C[orps]. Little or no damage. Our cooks have constructed a good brick oven in a side hill. Clay mud for Mortar. They now are making very palatable Dry apple pies charging 25¢ each, selling quickly all they can bake. The pies are the size of a common tin plate. 12 such can be baked at one heating. A Stirring arty duel came off late in the day. Few casualties.

July 26th Tuesday. Near Petersburg, Va. Pleasant, some warm. Campbell and Spicer were around to day with reading mater from the Christian Commission. Some little noise at the front. Sent my old & worn testament Home. Dress Parade at 5 p.m. Orders for Drill 2 hours each day. We are out of the Trenches. This could give us good practice. A few of us who Humbly try to live for God, had a season of Prayer in the evening. The Rebs threw over quite a number of shells during the night. Our Batts. Do not respond to any extent. Some reason at H[ea]d Q[uarte]rs has caused them to discontinue their discretionary fire unless a charge is made. The Batts. are to fire only upon direct orders from Div Hd Qrs. Our Batts. Had orders to return the fire during the night.

July 27th Wednesday. Near Petersburg, Va. Warm, some flying clouds. About Camp all day. Had a good head board marked & letters carved for Henry Balis' grave. Henry's Brother came to see me, and went with me, when I was ready to put up the board. Dress Parade at 5 p.m. Had another Prayer season during the evening. Very well enjoyed.

---

1 Newton Martin Curtis was later promoted to brigadier general and received a brevet promotion to major general. He led the brigade for much of the remaining service of the 117th New York.

2 Campbell was a member of the family who owned the Walcott & Campbell textile mills.

Capt. David B. Magill, Company D.
*Donald Wisnoski.*

July 28th Thursday. Near Petersburg, Va. Cool during morning, but the increasing directness of the suns rays dispelled the coolness ere the Meridian was reached. Wrote a letter to Minnie in the Morning. General Inspection at 3 p.m. Col. Curtis Inspector. Went into the Trenches at Dark. Campbell and Spicer were here again. Capt. Magill[1] & Myself went with them to visit Harry Balis' grave.

July 29th Friday. Near Petersburg, Va. Very warm through the day. Some Sharp Shooting, The "Johnnies" have thrown quite a number of shells. Our batteries have had orders to answer a little. At 4 p.m. rumors of Marching Orders began to agitate. They came more direct until we go the orders. Drew 1 days rations Soft Bread Sugar & Coffee. During Evening issued all rations on hand. 3 days. At 9 p.m. our Division moved from the Position we ha customarily occupied & went to the rear of the 9th A.C. as a support preparatory to the move which was in cause of arrangement. After getting into Column we laid down on our arms and got some sleep.

July 30th Saturday. Before Petersburg, Va.[2] At 3 a.m. all were roused from rest, and began moving forward quietly. At day light the mine beneath the fort was Exploded, throwing Earth, Cannon, rebs & every thing comprising the armament of the fort far upward & fell in a promiscuous manner those falling first being buried by those following. At the instant that the Earth began to tremble from the force of the mine, just then our Cannon opened the most terrific vollies of shot & shell. The rebs were completely surprised, & fled from their trenches in front. Our men followed, & occupied their first lines. Afterward they charged

[1] Born in Ireland, David B. Magill lived in New York Mills and was later breveted major for gallantry at Fort Fisher.

[2] The following description refers to the Battle of the Crater.

& took others. Our Cannon in the mean time spitting forth in tremendous amounts of iron hail. The success of the explosion & the early success filled our men with the most exhillerent enthusiasm. At this time squads of rebs began to be guard[ed] to the rear. Some of them were wounded. They all were dirty enough, as their tumbling about at the Explosion was not at all conducive to neatness of appearance. Many of them were without hats, coat shoes or stockings. Our Div began again to move father forward, & to get into the trenches. Some of our wounded were being [brought] back. As the morn advanced other charges were made which did not terminate so favorably for our forces. The negroes when they charged forward were met with such exasperation and desperation by the rebs that the Negroes could or did not withstand them, & were forced back. This circumstance was disadvantageous to us. Many of the white troops were panic struck, & behaved like anything but soldiers. Shall a Soldier quake with fear, & falter when the fighting rages? Never. Unless completely overwhelmed & driven a soldier should manfully combat & when driven should by every device & power possible harass the enemy & take care of the flying or better retreating forces. Many our regt Could see over the field in which this transpired & were provoked to see white Men & Negroes come rushing back without gun or hats in many cases. These called out to us to fall back, but Knowing that there was no necessity for this piece of work we filled the trench full, but some of them so wildly excited & bent on getting back, clambored up the sides of the trenches & pushed slid backward, foolish. This lasted but a short time, as the better class of men & there are many, Officers & Privates, who used example & command to counteract such disgraceful conduct. I was disgusted, mad roilly, especially as it gave the Johnnies the opportunity to shout & scream out their pleasure at our panic, & the rebs did howl considerably, well they might but the Johnnies did not venture to crowd us as they knew as well as we that if they venture against our works that they are moving against a yawning grave in which secession would be speedily thrown, should they practice assaulting us, instead of the necessity which has made us the assaulting party. The contest lulled, Our men holding their own position defiantly, & tenaciously holding to a portion gained from the rebs including the defunct fort. Our Div Moved back to the rear, forming column in the woods. This was after M. We rested here, cooked coffee, got dinner. Drew whiskey as usual for the men. Mail came. Got letter from Hannah. The Excessive heat of the day had very much fatigued us all. The rebs about this time receive large reinforcements, & crowding forward, our men were forced mainly to let go their hold on what they had gained from them. This was the end of the ernest work of that day. Our men all felt mad at the manner of its termination.

We could have forced the rebs out of their Position, as 30,000 of our men were not engaged at all, but I suppose the Commanding Officers came to the conclusion that to risk such a sweeping Engagement was not in keeping with the welfare of the army. If the Engagement had resulted in a hard fought battle between Infantry, thousands of men would have fallen from sun stroke. Some did as it was. Later in the p.m. Our div moved back toward our former locality, going, at night, into the trenches. The 117th taking position in the 2nd line. Quiet during night. Slept most soundly.

July 31st Sunday. Before Petersburg, Va. Three were slightly wounded yesterday by long shots. Early day saw us moving toward City Point, destined ultimately for Washington to operate against the Northward inclined Johnnies. The sun as day advanced increased its rays to their acutest intensity. We marched slow but the flood of sun heat which poured down upon us, fatigued us greatly. As we neared the locality of City Point, we recd orders to move to Bermuda Hundred Fortifications. Turning hitherward we crawled along under the searing heat. Many began to "fall out" as we neared the Appomattox. I was with the regt rear guard. The river runs between high bluffs on either side & steep hills were hard to descend & climb, this with the heat while crossing on the Pontoons, came near completely "whipping" the majority of our men. Large numbers of our men "fell out" in ascending the hill, after crossing the Bridge, & many more on top of the hill, but the column did not move but a trifle farther, the majority of them worked their way up. Here we stacked arms and rested till toward evening when we fell in & moved to the fortifications. Got into Column, stacked arms, laid down, & slept soundly. This day's march came the nearest to "using me up" than any since coming into the army.

---

*Shortly after midnight, on the early morning of July 31, notification arrived that the regiment was about to be rotated out of the front line. During the long weeks it served in the Petersburg trenches some of its soldiers died from sharpshooters and shell fragments, more fell victim to the deplorable conditions and diseases. All of them suffered through heat, humidity, vermin, dirt, rain, mud, and other hardships. It was an experience those who survived would never forget.*

# Bermuda Hundred

## August 1-27, 1864

*On July 31, 1864, the regiment received orders to return to the Bermuda Hundred lines where the men had an opportunity for a few days' rest. During this interval, Gen. Grant determined to continue his attempt to cut off Petersburg from the south. As part of the move, Grant ordered Gen. Butler to send most of his troops north to threaten Richmond so the Confederates could not shift troops to meet his new move south of Petersburg. Curtis' brigade was among those left behind to hold the Bermuda Hundred lines. Fortunately, their front remained relatively peaceful, the pickets on both sides coming to an understanding lessened the danger from sharpshooters during the day. The lone attack on the regiment's lines was met and thrown back.*[1]

---

Aug. 1st Monday. Bermuda Hundred. Change locality a little so as to allow the Div to get into proper position. Began clearing for camp. Worked as well as the heat and our Physical strength allow. Wrote to Hannah.

Aug. 2nd Tuesday. Bermuda Hundred, Va. At work rigging Quarters. Men Detailed to make Officers Qrs comfortable by building bough shades over tents, & after this is done they can, if a mind, make themselves comfortable. Officers make very little calculation on producing their own comfort. Arrogance & Imposition are often perceivable, & very much in those Officers who have been promoted from the rank of Enlisted men.

Aug. 3rd Wednesday. Bermuda Hundred, Va. Mail came at Early morn. John recd small box from Home, containing soft hat for him, Pack of socks for me, & sundry other articles. Commenced digging well, busy cleaning ground. Sutler came to regt yesterday giving our checks. He has small brass & Pewter pieces for a circulating media instead of the tickets in former use. The men are getting orders to the amount of $5.00 on average.

Aug. 4th Thursday. Bermuda Hundred, Va. Our Camp is not regular enough to suit our "Superior Officers." The tents of some of the Cos.

---

1 Longacre, *Amateurs*, 195-96; Mowris, *History*, 128-29.

Fortifications in the Bermuda Hundred lines near Point of Rocks. Note how stakes are pounded into the ground, logs piled between them, and then covered with dirt to make fieldworks. Note also how the smaller trees have been cleared for a field of fire and the earth has been banked to protect the camp in the background. *Library of Congress.*

---

are being moved, coming upon a more Exact geometrical line. I wish the officers were forced to clear the ground upon which we are to camp, & they would not be so particular as to the Exactness of right lines. Especially while campaigning. I bought a pair of worsted shirt[s] from the Sutler, paying $10.00 for them. Drew 5 days rations to day. A large detail 300 men were set at work during the cool of the evening in clearing up a considerable space to the rear of the camp. "D" Co. was ordered to repitch their tents. The variation from a right line in the distance of our co. was 5 [?] feet. Some of the tents did not get moved more than as many inches. Much more exacture then sense.

Aug. 5th Friday. Bermuda Hundred, Va. Hot as usual. At work on the well. Men busy rigging up camp. Pickets Detain sent from our Brig to night. Terriffic cannonading at the Petersburg front commencing about 6 p.m. & lasting till sunlight had passed. The evening was cool. Got letter from my Minnie. Wrote answer.

Aug. 6th Saturday. Bermuda Hundred, Va. Very hot. Reports of last Evening's work at Petersburg are coming in. Hope they will prove true. Rumors. The rebs undermined the Batt. Near which our "Div" had Position. Our men received information of the fact that they were purposing to blow it up last evening, moved all their guns & material to the rear of that hill. The Rebs set the fuse to the mine on fire, the thing went off, & then the rebs charged. Our Troops were in position to receive them nicely. Results of the Rebel Scheme 3,000 reb prisoners in our hands, and many of them 1,500 Killed & wounded. Meager losses of Killed & wounded on our side. Drew from Commissary rations of Mackerel & Codfish which are in lieu of salt meat one day.

Aug. 7th Sunday. Bermuda Hundred, Va. Hot as usual. Company Inspection at 7 a.m. About Camp all day. Went to a Prayer meeting lead by the Chaplain of the 112th N.Y.V. They are, I should judge, accustomed to private Communions with our Blessed Master, as not one of them spoke in a tone above a mumbling jargon, wholly unintelligible to a man who might sit 3 paces from him.

Aug. 8th Monday. Bermuda Hundred, Va. Hot. Drew fresh meat, covered with wee fly blows. Cut the best of it into steaks & gave it out. The rest, more than half, I burned. Also got fresh bread & whiskey. Wrote to My Mother. Picket detail from our Brig to night. Had a good wash in Cool water taken from our well.

Aug. 9th Tuesday. Bermuda Hundred, Va. Hot as usual. S. R. Campbell and J. B. Spicer came to see us this morn. They are going Home tomorrow: so they say. Wrote to Minnie, acknowledging the recei[p]t of a Testament & some papers. Our Camp is improving under the thorough scraping and sweeping. All smooth in this location.

Aug. 10th Wednesday. Bermuda Hundred, Va. Hot. 8 of us were occupying the 4 tents on the left of the co. Made a Pine bough shade over out tents. This has a positive effect in making this special locality much more endurable. Got letter from Minnie to day. Was selected as "left" General guide of the Battalion. Ordered for Co. roll calls were promulgated to day. Reveillie, breakfast, dinner, supper, Retreat and Tattoo.

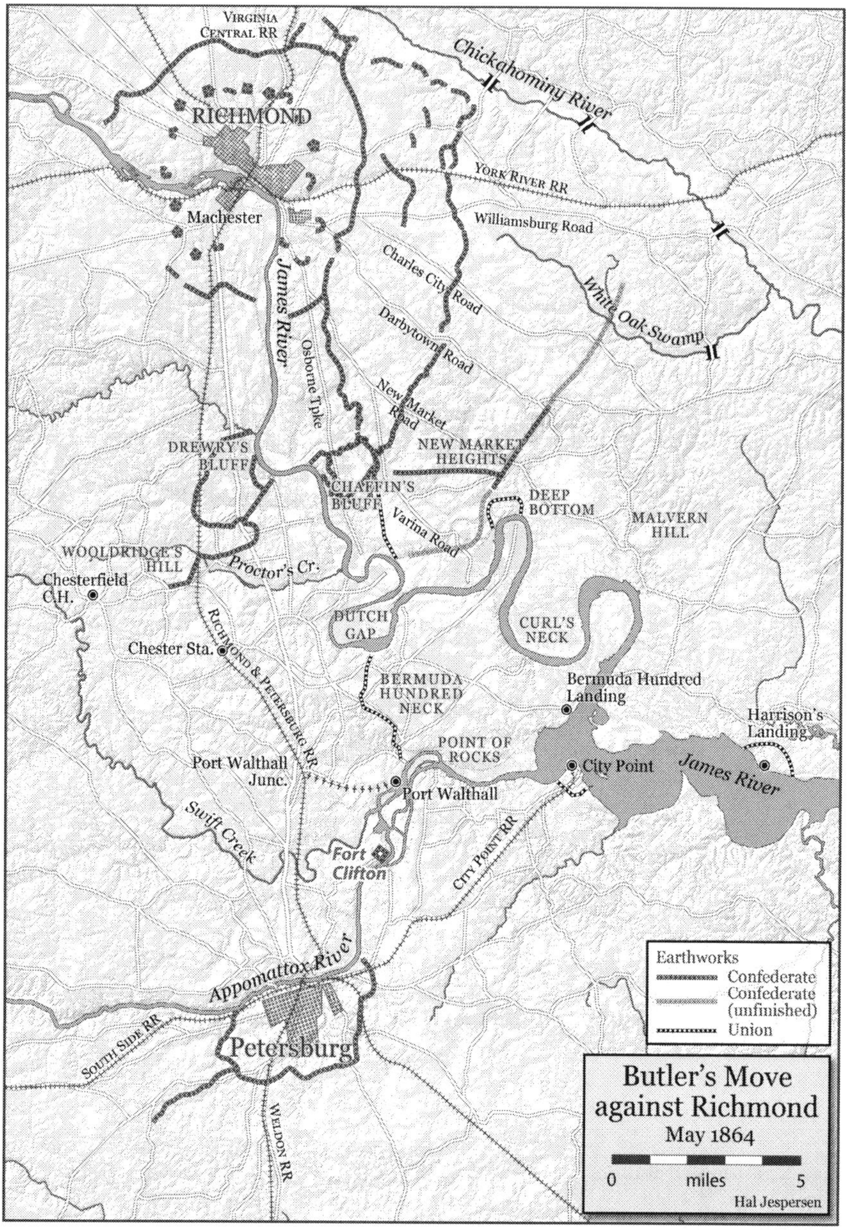


The Bermuda Hundred area between Richmond and Petersburg including the areas where the regiment was later active at New Market Heights and the Darbytown Road. *Hal Jespersen.*

Also drill by squads, Co. Inspection each morn. Had a good sing in the evening with Green, Brown & Kimble. Coolish evening. New Moon, bring thoughts of home.

Aug. 11th Thursday. Bermuda Hundred, Va. Roused at Half past 3. Ordered to fall in Under Arms. Went from Camp to the Breastworks lying till after sunrise came. Back to camp. Did some washing after Guard Mounting. Our Boys are using their time in making and cooking dumplings and puddings.

Aug. 12th Friday. Bermuda Hundred, Va. Reveille at 5 a.m. Got additional brush for my shade this morn. Day passed in having "roll calls." The men howled very much. Col. Curtis was about and gave out stringent orders against the howling. It has ceased. This is a wholesome bit of discipline. Wrote to Hannah. "Roll Calls" were attended 6 times. Dress Parade at Sun Set.

Aug. 13th Saturday. Bermuda Hundred, Va. Weather unchanged. Reveille at Sun rise. Early in the day Orders came to have three days cooked rations in haversacks, & ready to march, by 5 p.m. One days rations were issued to make up the compliment. Were busy with the rations most of the day. All ready long before the time. At 5 p.m. Every man, & most of the non-com officers, made a large detail for Picket. This was the case throughout the Brig. Our Camp is nearly deserted. The Sick & a few others are all that are about. Got letter from Hannah at Early morn. Mail is Early. Wrote to Hannah. The sick who are unable to walk were carried by the Ambulance trains, going at 8 p.m. The few left behind laid down on bunks to sleep.

Aug. 14th Sunday. Bermuda Hundred, Va. At 3 a.m. all well men remaining in camp, were got up, "fell in," & went to Battery No. 1, as a reserve. Laid along the parapet till 8 a.m., then taking possession of the shades of a vacated camp of a position in the 1st Div. Hot during the day. Much thunder during the latter portion of the p.m. Some little rain; put tents up on the polls used by the former occupants of the shades. There is bearly a corporal's guard within the intrenchments to day. Some fighting on the North bank of the James River. The fighting on the North of the James is between the 10th A.C. & the Johnnies. Our Brig is all the Troops remaining here, and are doing the Picketing. At 4 p.m. Every well man in camp, excepting cooks, recd orders to go on Picket. Orderly Sergts Pioneer Corps & all were out. We Made considerable numbers.

Aug. 15th Monday. Bermuda Hundred, Va. The Johnnies were very quiet last night. Their videtts were somewhat close to ours. During to day, some of our men along the line had converse with some rebs. Some few were seen in our immediate front but distant. The Rebs were much more charry about showing themselves than our men are whenever our lines & the enemies are in close proximity, so that our men can [not?] talk with the Grays, they have an officer on duty. In conversation with our men the rebs gave ideas that are not at all favorable to the Southern notion of Independence. Hence the care the reb commanders to prevent communications. We were release at 6 p.m. Came to camp. Had good night's rest. Some stir in camp. News from our Corps is cheering. The report says two lines of rebs works and one battery are captured & that our men have driven the Johnnies. Thunder & some rain latter part of p.m.

Aug. 16th Tuesday. Bermuda Hundred, Va. Cloudy, a misty rain at early a.m. Drew soft Bread & fresh meat. Those on Picket draw whiskey ration: ½ gill. The detail for Picket from our regt was small (100). We ranged ourselves more regularly during the early evening. Every body was exceedingly busy for two hours. Some more cannonading during the p.m. on North side. No absolutely reliable news as yet.

Aug. 17th Wednesday. Bermuda Hundred, Va. Hot during the early part of the day, a strong breeze. After part was the scene of terrible thunder and lightening but little rain. This section gets up some tremendous demonstrations of the upper artillery but brings very little of rain. Our Gun boats batteries had a considerable duel with the Johnnies Gun boats & batteries during at 4 p.m. The Grays made some good shots also did our men. Our arty threw the most Shells.

Aug. 18th Thursday. Bermuda Hundred, Va. Thunder headed clouds hang low this morn. No breeze. The air is very hot and sultry. Heavy cannonading from the direction Petersburg at early morn. A few shots have been thrown on the James. Had General Inspection at 7 a.m. Beebee of Curtis' staff inspected us. A Private of "C" Co. sent out by a Staff Officer to exchange papers with the Johnnies was "snatched" by them. They seem to be absolutely (at times) adverse to exchanging papers. Whenever they do exchange, they tear off part of their papers as though it contained information which they cannot be pleased or comforted by our Knowing. A heavy shower During a.m. Also much sharp lightening & thunder. William Wright of "H" Co. was Killed by lightening. He was on Picket. The current of electricity passed along his gun into his left foot & off by his shoulder tearing his shoe and stocking slightly. His body was not torn any. His flesh when brought to camp was a bluish purple. He was lying down sleeping at the time feet toward the tree, gun resting against it. A large man 180# or more.[1] Some heavy firing at Petersburg during p.m. Wrote to Sister Annie. More fighting on the North Side at evening. Very quiet here in our front. Some cracking on the river.

Aug. 19th Friday. Bermuda Hundred, Va. Heavy Clouds all day. Copious Shower from 4 p.m. till 7 p.m. Heavy work at Petersburg to day, very rapid firing late in the p.m. Only a few shots exchanged between the river and the batteries. Wrote to my wife to day. Wrote letter for S. D. Powers. John recd letters from Home.

Aug 20th Saturday. Bermuda Hundred, Va. Weather same as yesterday. Heavy firing in the direction of Petersburg during last night. Drew 5 day's rations. Recd letter from Minnie this morn. Rev. Jones,[2] our new chaplain, arrived to day. He is a man about 27 years of age.

Aug. 21st Sunday. Bermuda Hundred, Va. The 1st Div came back at 2 a.m. The 6th C[onn]. Vols. Came to their old camp, the one we were occupying. We got up, & packing moved to our own camp. Cloudy during the day. Busy fixing tent all forenoon. Heavy Cannonading in the direction of Petersburg. Had sing to night. Orders Came to be ready to march at 7:30 p.m. Light marching orders, one days rations. Laid down to wait. The move was not made till early morn.

Aug. 22nd Monday. Bermuda Hundred, Va. At 2 a.m. we "fell in" and moved to the left of these fortifications, passed by the flank of works. Soon reached our Picket line, were making forward to get into column. The prospect was that we were preparing for a desperate charge. But orders came from the rear countermanding the movement. "About face, forward March," and we started camp ward. It turns out that our move was a counter movement to an anticipated reb movement on our forces holding the Weldon R. Road. The Rebs did not move, & we did not prosecute ours. Got back to camp at daylight. Cloudy, & cool all day, rainy during evening. Picket went out to night. 100 men.

---

[1] William B. Wright was a laborer from Vienna, NY.

[2] John D. Jones was a Hamilton College graduate with prior service as a private in the 17th Connecticut Infantry. He joined the regiment at Bermuda Hundred on June 22, 1864.

Lt. John Thomas, Company C.
*US Army Heritage Education Center.*

Aug. 23rd Tuesday. Bermuda Hundred, Va. Coolish, some sun shine. Were wake at 4 a.m. "Fell in" and went to fortifications, returned at daylight. Wrote to wife. Also sent her my filled Diary. Made out three sets of Papers for Lt. Thomas.[1] Dress Parade at 6:30 p.m. Scribbled during the Evening.

Aug. 24th Wednesday. Bermuda Hundred, Va. Clear sky & hot sun. Went to the fortifications at 4 a.m., returned at daylight. Orders for Gen. Inspection. Got ready 4 p.m. Orders to move, conflicting ideas. Large detail set out on Picket. Orders at 6 p.m. to pack everything. Did so. Marched to fortifications. Stacked arms. All indefiniteness as to destination or object. Darkness coming on laid down. 1st Div. moved Petersburg way. Slept behind stacks of Arms. Cooks drew 3 days rations.

Aug. 25th Thursday. Bermuda Hundred, Va. Just before 4 a.m. I waked up, selected 6 men to assist in getting rations up to the Co. Turned to go, & Crack went a musket on the picket line. Another, then yelling & a volley. Sharp firing. I remained at the Co. till the firing was mostly over. Went after rations. A little after daylight the firing was renewed & with more vigor. The rebs, who had succeeded in the surprise of the Early morning getting possession of our Picket line partly, were driven out, & rushed [?] hastily for there old Pits. Our Men had charged, retaking our line also a number of reb Prisoners. Various rumors were flying amongst us all day. Our men lost some few in Prisoners. A very few in wounded. One or two dead. Our Pickets coming in to night make the following statements. "We had notice of a stir among the rebs. Were wide awake. But the section occupied by us, was dense woods preventing us from seeing any object. The Rebs cautiously moved one at a time

[1] John T. Thomas was a stonecutter from Trenton. He was later killed in the assault on Fort Fisher, NC, January 15, 1865.

Capt. Augustus Erwin, Company K.
*Donald Wisnoski.*

to their vidette posts. A short rapid roll from a drum, & instantly the rebs began to rush forward as vigorously as their leaders could urge the men. In some cases the men seemed disinclined to move as there was much swearing & urging on the part of the leaders. Still under the darkness they rushed forward not firing much as they came, but calling to our men to surrender. We fell back, firing, to another Pit. When the Johnnies saw we had no inclination to surrender, they fired many shots but fortunately they went high." Two were slightly wounded. In this the Grays succeeded in capturing a few out of our regt. 2 from our Co. E. E. Tubbs & Geo. Russell.[1] As soon as daylight came we were reinforced by 100 men, forming we charged forward, taking our old lines, a number of rebs, guns, & a quantity of spades & shovels they had brought with them to reverse & strengthen our pits. These pits were first built by the Johnnies, captured & reversed, & have been held a long time by our forces. The Grays got out of our pits in hasty confusion when we charged. The Johnnies were fortunate in getting a large number of haversacks, canteens, & some knapsacks. These they "snatched" eagerly. The Rebs suffered much more in Killed wounded & prisoners than we did. Our men buried 7 and some were wounded. Three lay between our & the reb lines all day. Our men carried them water although one of them was nearer the Reb line than ours. After dark he crawled into our lines. Farther on the left of the line a reb Officer was very officially & magnanimously inviting our men to desert saying we wont hurt you. This officer felt a sharp severe pain in his breast, & dropped instantly. A "Yank" had sent a piece of lead through him. A poor remuneration for his generosity.

On the right where most of the excitement occurred, the firing ceased & all was quiet during the day. In the center there was a little

[1] Edwin E. Tubbs was a Utica machinist. George R. Russell was a farmer from Sangerfield. Both men died in the Confederate prison at Salisbury, NC, in November 1864.

popping during the day. Our arty shelled the Rebs severely. They very timidly replied. All became quiet at eve. Rain at 7 p.m. for an hour. Our regt laying at the fortifications. Capt. Erwin[1] was wounded in the arm, not severe. A man in "G" Co. was hit by a bullet, causing a contusion on the back of the head, not serious. We are prepared for a moments motion.

Aug. 26th Friday. Bermuda Hundred, Va. Sun rose clear, warm. All quiet, no disturbance from the Johnnies. They must feel some what chagrined at their failure of yesterday. From the stories of the boys, it is evident that they did some "tall tumbling" to get away from the Johnnies when they sprang upon them so suddenly. Some of our boys lost their caps in the melee as they ran through the brush. Rain at 8 p.m. Some wet all night. Laid on three poles, having my rubber blanket thrown over me, Kept dry.

---

*While Fairhead protected himself from the rain, orders arrived for yet another move. The following day the men fell in to begin the march back to the dirty, diseased, and dangerous trenches outside Petersburg.*

---

[1] Capt. Augustus M. Erwin was from Rome.

# Petersburg

## August 28-September 27, 1864

*On August 27, the regiment once again began a move, headed back to the trenches at Petersburg. Crossing the Appomattox, the men filed into the fortifications to relieve the 29th Regiment, U.S. Colored Troops, the right flank of the regiment resting on the Appomattox River just north of their previous position in the line and only an estimated quarter of a mile from the city's streets. Being so close to Petersburg, this was a very dangerous section of the siege works with active sharpshooters and shelling. Despite the constant hazards of the front lines, occasional contact still took place between the belligerents where the 117th New York faced Confederates across a narrow place along the river that precluded an immediate infantry attack. Intrepid soldiers occasionally exchanged newspapers or traded coffee for tobacco.*

*In early September news arrived of the capture of Atlanta, followed by positive reports from the Shenandoah Valley and other Union successes that heartened the men, strengthening their morale in the face of the daily drudgery and anxiety in the trenches. Brief rotations brought them to the rear for rest, but usually still within range of Confederate artillery.*

---

Aug. 27th Saturday. Bermuda Hundred, Va. Cloudy & cool during early day. Moved toward Petersburg, reached the rear of our forces there at noon. Laid on the plain till dark, then moving into the trenches on the "right." The right of the regt resting on the Appomattox river. We releaved the 29th U.S.C. regt.[1]

Aug. 28th Sunday. Some little arty firing during the night. Other wise all quiet near us. The day was some warm. I made Bomb proof during last night. They also make excellent sun proof. The Johnnies kept up their musketry poping, though they do no damage, nor do we pay any attention to them.

Aug. 29th Monday. Weather as yesterday. In pits all day. Made ourselves comfortable. At 5 p.m. a severe arty fight commenced. The Johnnies sent a few shot directly into the Trenches occupied by our Regt. It

---

[1] U.S.C. refers to United States Colored Troops, African-American regiments that were led by White officers.

is surprising that they did so little injury. One man lost his foot, three others recd slight bruises. At 8 p.m. all our Batts opened on the Rebs and Petersburg. This was a terrific cannonade. 10 shells were counted going into the City at once, and there were others flying away to Reb Batts. This lasted two hours. The reply of the rebs was like a squirt gun stream. We could hear the crash of the shell as they would tear through buildings. The Dogs at least were frantic. Got letters from Hannah. Commenced answer. Commenced "Muster Rolls" to day. Made out the first copy. Wrote a letter to E. E. Tubbs Mother regarding his capture.

Aug. 30th Tuesday. As yesterday, worked all day on "Muster Rolls." John G. Glazier[1] was Mustered as 2nd Lt. in Co. "D" to day. He was 1st Sergt of Co. "A." Arty very quiet at evening a few shots were fired. Musket poping mostly stopped. Wrote to Hannah.

Aug. 31st Wednesday. Pleasant weather. Gave the finishing touches to the "Muster rolls" this morn.[2] Were "mustered" in the pits. Got letter from Hannah. Were releaved from the trenches by the 112th & 3rd N.Y.V. Our Boys had some haranguing with the Johnnies just before we came out of the trenches.

Sept. 1st Thursday. Bright & breezy. Wrote to Parents. Some arty at 10 a.m. Gave our muster Rolls their finishing touch. The Men made themselves comfortable. Many of them went to the fields in the rear to get green corn, apples, mellons and what other eatable green vegetables could be found. They got considerable.

Sept. 2nd Friday. Weather as yesterday. Some little arty. Wrote to Hannah. Also wrote to Capt Storrs Barrows. After we had lain down at 9 p.m. yelling was heard from the front, our boys pricked up their ears, thinking that a charge was being started. A little popping was all that could be heard. The noise of voices was continued, developing into regular cheers. We made up our minds that it was not a charge. The Hurrahs went the whole length of the line. After waiting with much anxiety we learned the cause of the cheering. A report from Corps Hd. Qrs., said that Atlanta was captured by Genl Sherman. The Cheering was electrical. Some of the Johnnies Fiendishly yelled as our cheering com-

---

1 John G. Glazier was a tinsmith from Verona.

2 As company sergeant, Fairhead was responsible for keeping the muster rolls current each month.

menced & continued. One of our 4 mortar Batts gave the Rebs a full salute twice. The Johnnies answered a little. A couple of our Brass Bands gave us the National airs in good style. Our boys feel jubilant. Success pleases. Reverse should give additional firmness.

Sept. 3rd Saturday. Air cool and stirring with much cloud. More information about Atlanta and Mobile, both reported captured. Our Batts had a set to with the Jacks at 1:30 p.m. Wrote note & sent a few verses to N. C. Advocate. Went into the pits at 9 p.m. A vigorous cannonade prevented us from going in at an earlier hour. During the night there was much quiet.

Sept. 4th Sunday. Air stirring, warm. Worked some of the day making bomb proof. Got a square hole dug & most of the timber ready for covering. Put it off till Monday.

Sept. 5th Monday. Air & sun hot. Last night at 12 M. all our batteries opened on the Johnnies. Our arty men gave three cheers as they conducted their work. This was strange proceeding. The Johnnies were slow to open, but finally did with considerable. The lurid flash of the guns, the incessant roar of the discharges, the lightening speed of the shot whose pathway was easily traced by the burning fuse, all accompanied with the rushing roar, made the scene more terribly sublime. I stood in the excavation for my proof watching the course of the shells. The Johnny Batt. on our flank, across the river, was very actively engaged in throwing shell along our lines. A number stuck immediately in our trenches, pieces flying very freely. It was strange that so few were injured. Only three or four, slightly. A few guns were shattered. As I stood watching this, observed a Johnny shell with its eye of fire coming directly toward me. As it came close I dropped down in the hole and listened intently for the strike & explosion. I waited but an instant. Thump it went into the fresh dirt that had been thrown out, & in an instant more crak went the shell. A shower of dirt fell over me, I waited a moment more to let pieces of shell fall wherever they would. None hit me. I got up, shook myself, could See nothing as a cloud of smoke had shut out every thing from view. I called out "all right." "I'll stay here as a shell never comes in the same place twice." This shell struck about 6 feet from me. At daylight I picked up 5 pieces of shell which had fallen on the new dirt. The shell was a twenty pounder (conical) shot from a rifled gun. Finished my [bomb]proof. Tis a good one. Got letter from home Hannah. Rain commenced falling at 10 p.m. My bomb proof is perfectly water proof also. We learned that the firing during the night was a jubilation over the victory at Atlanta each battery firing a national

A "bombproof" along the Petersburg fortifications. These were often made by constructing a log enclosure and then shoveling enough dirt on top to absorb the impact of shells and fragments. Some had wooden covers that could be placed over the opening as necessary. *Library of Congress.*

---

Salute. The Johnnies must have thought strange of this affair.

Sept. 6th Tuesday. Lowery[1] & showery. Drew three days rations. Sergt Clarke ordered under arrest for indiscreet language to superior officers during last night. He referred to the short comings of said Officers as regards their duty & plead ignorance of duty himself. Were releaved from the pits at 7 p.m. Went back to the ravine and put up tents. I went in with James Finch.[2] Damp all night.

Sept. 7th Wednesday. Some lowery and cool. Went to Brig and Div wagon Trains to see our Boys. Had dinner with R. Johnson[3] who cooks for the Teamsters. Baked Beans. They were excellent.

---

[1] Gloomy.

[2] James Finch, a carpenter from New York Mills.

[3] Robert Johnson, a baker from New York Mills.

James Finch, Company D. Note the letter D and the 117 on his cap to the right. *Oneida County History Center.*

Sept. 8th Thursday. Warmer, damp & chilly all night. About 25 per cent of our men are ailing with Fever & ague, and Kindred ills. Wrote to my wife. Did considerable working for Lt. Thomas, Com'd'g Co. Was busy all day. Drew rations during afternoon.

Sept. 9th Friday. Sun clear, breeze coolish. Made out Requisition of clothing for October, Co. "D." The Johnnies and our Batts had something of a round at the middle of the p.m. The Johns have a gun in a new position, bearing upon our old 3.30 pounder Batt. This new Johnny Development will not amount to any large sum. Our Batts can soon drive the gunners from their posts. They are erecting some new Earth works on our right flank. Uncle Sam has as many guns as one eyed Jeff. Went into trenches at dusk. As soon as we got in & some what settled order from Genl Grant came that we should give three cheers over success in Georgia. Our men gave them with zest. The Johnnies Howled some when we returned the cheers. It is Energizing to hear cheers run along our lines reechoed by those lying in the rear, & taken up by the brass bands. I after ward worked in charge of a squad repairing and increasing abbatis in front of pits.

Sept. 10th Saturday. Sultry. Some little arty. Drew rations, got three letters. Hannah, Annie & Mother. Wrote to Mother. Good news from Sherman, not so definite as we wish, but reliable. An Agent in the U.S.C.C.[1] just gave me a copy of The Gospel by St. Mathew. I did it up and sent it to my wife.

Sept. 11th Sunday. Comparatively quiet all day. Commenced letter to Hannah. In Trenches. Chaplain Jones[2] was through the Trenches with papers. Gave me two.

---

[1] United States Chaplain Corps.

[2] John D. Jones, a graduate of Hamilton College, joined June, 22, 1864, after service as a private in the 17th Connecticut.

Sept. 12th Monday. Rather rapid musketry popping. I sent three shots at a Rebel port hole. Came out of Trenches at dusk, were releaved by the 112th. Bright moon light nights. Got letter from wife.

Sept. 13th Tuesday. Showery, finished letter to Hannah. Was busy with Co. papers. No disturbance.

Sept. 14th Wednesday. Warm. A heavy arty duel was commenced by our Batts. Petersburg recd a number of Peace Propositions. One of our flank Batts severely handled a Reb Batt across the Appomattox. The Johnnies were much surprised if not hurt, as they asked "what sort of arty we had.["] The Batty had been changed and this was the first engagement. The Johnnys were very busy preparing their works after the engagement. <u>Wrote</u> to <u>Hannah</u>; also to <u>Mrs</u>. <u>Whetmore</u>. Was busy at Company books and papers. Wrote a note to H. W. & J. P. Garrett, 97th N. Y. Vols., 5th A.C.

Sept. 15th Thursday. Hot. Finished up a set of papers regarding ordnance. Quiet at the front up to this time. Wrote to Brother James. Went into the Trenches at dusk. Very quiet & the most beautiful moon light. Sad that the moon in its refulgent beauty and purity should see such hideous sight as two immense armies confronting each other with deadly weapons.

Sept. 16th Friday. Unobscured sky, sun & Time sped by as pleasantly as possibly. Wrote to C. A. Boundy, Boston, Mass.

Sept. 17th Saturday. Weather unchanged. In the Trenches. Johnnies were very quiet during the day. There was a tacit agreement not to fire. Just at sun set they fired a little. They say that their officers command them to do so. Got letter from wife. Well & cheerful. Wrote to wife. A few cannon shot only, passed between the Batts. Our Fatigue Parties are at work on the new Fort.

Sept. 18th Sunday. Some little cloudy. Read in my testament some in the morning. Drew rations. Soft bread at evening. Were releaved at 6:30 p.m. Came to rear & put up tents.

Sept. 19th Monday. Coolish but dry. Quiet. Regtl Inspection at 9 a.m. Our Blankets which were packed and stored last spring when the Campaign was opening, came this a.m. The men are some what pleased to receive then as the nights begin to be cool. Some who were with us then, but are now sleeping in a soldier's grave, were brought to mind by

Although officers frowned on the practice, when they were not present and a temporary truce could be arranged, Union and Confederate soldiers would meet between the lines to exchange newspapers or trade items like Northern coffee for Southern tobacco as seen in this Edwin Forbes sketch. *Library of Congress.*

unpacking their blankets. Peace & honor to their memory. We are spared for future work. May we do it as well as they. If we fall we shall fill graves which will ever be honored, unless it may be by them who inaugurated this war and men who are steeped in sympathy with those Traitors. A few shots were exchanged with the Johnny arty at evening. Wrote to J. H. French.

Sept. 20th Tuesday. Fair weather. Genl Inspection at 8 a.m. Co. turned out in good order. Muskets bright, brasses clean, straps & accoutrements blacked & polished, shoes the same, faces washed & shaved. The remainder of the day, I occupied in drawing rations & other business. Went with rations to three of our boys who are at the convalescent camp a mile to the rear. At dusk the first news of victory in the Shenandoah valley came to the Troops. Strong & loud cheering went the whole length of our line.

September 21st Wednesday. Heavy fog at morn but fair weather through the day. All our Batteries opened at sunrise, each gun throwing shot or shell over into Dixie in honor of the victory. The roar was almost deafening but quieted down by breakfast time & remained so all day. I cant vouch that the Johnnies like this way of celebrating our victories. Our regt went into the Pits to night. After getting into our usual positions, we got orders to cheer again, as more news of victory came and we did cheer most lustily. The Jacks were sullen this time as they did not get up a counter noise as they have done heretofore. Got letter & box of wedding cake from Home this morn. Some Plumb pudding also.

Sept. 22nd Thursday. Cloudy with a trifle of rain. The Johnnies are considerable quiet, but they pop away at times. Once in a while the Rebs will throw a shot or two at our teams as they pass to the new fort we are constructing. This they have practiced some time, but with no damage as yet. They have made a number of close shots, but a miss is as free from injury as though the ball went a mile out of direct course.

Sept. 23rd Friday. As yesterday. Affairs in our front about the same as on 22nd. Capt. Warr[1] Exposed himself this morn & was shot through the fleshy part of the thigh, some what severely, but not considered dangerous. Got letter from wife. <u>Wrote</u> to <u>wife</u>. Also, <u>wrote</u> to <u>Father</u>.

Sept. 24th Saturday. Some wet. Trenches are muddy. Later part of day orders came to be in readiness to move. I went at work [on] the rations & soon had the 4 days rations issued to the men. Their haversacks were well distended. We waited expectingly till 12 midnight, when the 1st Div of the 2nd Corps releaved us. We formed Regimental line & marched to the rear, passing by 10th Corps Hd Qrs and then Halted. Stacked arms, took out our rubber & woolen blankets, & with tent made our beds. We were two hours & ½ in coming from the Pits to this place, a distance of about 1 ½ miles.

Sept. 25th Sunday. Sun clear, coolish, put up our tents. Drew two days more rations. Boys used up the time in camp life. Corpl Green, Corpl King & Pvt. Kendall,[2] all of "G" Co, & myself went to the tent of a friend of their and had pleasant time singing during the evening.

---

1 Edward Warr was captain in Oneida County's 14th New York before joining the 117th. Fairhead described him as "firm in spirit, amiable & manly. Every fiber heroic."

2 George W. Green was a clerk from Paris. George H. King was a farmer from Rome. Austin J. Kendall was a merchant from Augusta. King was later killed at Fort Fisher in January 1865.

Capt. Edward Warr, Company C.
*Oneida County History Center.*

Corp. George King, Company G.
*Army Heritage Education Center.*

---

Sept. 26th Monday. Weather as yesterday. Still in same position. Men generally cleaned their guns today. Speculation is as usual on the eve of any change, very rife as to future operations. We are resting nicely. There is a strangeness of impression while moving about freely & without any whistling bullets, or schreeking shell. One has a sense of releaf as he moves about not circumscribed by the running of the Trenches, and feels as tho he was treading upon the surface of the Earth instead of being confined by banks of Earth, shutting out the rest of the section. Got letter from Mother stating the facts resulting in the Death of Mr. Bayne, Killed in the mill. He was badly mangled. His garments were completely torn from him, leaving not so much as "a wrist band" upon him.

Sept. 27th Tuesday. Warm. Order for drill 4 hours a day. Drilled 2 hours in the a.m. Moved & regularly arranged camp in the p.m. Fixed up out tent. Many "hearsays" regarding pay. It would be acceptable at any time. We have well earned our money during the last 6 months. And, yet, we are willing to fight on until it comes.

---

*To distract Confederate leaders from his planned movement south and west of Petersburg, in mid-September Gen. Grant determined to mount a demonstration toward Richmond using a portion of Gen. But-*

*ler's force. When orders arrived in the 117th New York to reduce baggage to a minimum, everyone realized another movement was pending. Speculation swept the camp, some imagining a transfer by ship to North Carolina, others to Hilton Head in South Carolina, while still others prophesied an advance to open the Norfolk & Petersburg Railroad. Sergeant Hermon Clarke from Waterville confided his hope in a letter home: "Where we shall turn up I don't know—I hope a safe distance from the Army of the Potomac."*[1]

---

[1] Letter, Clarke to brother, Sept. 25, 1864, in Harry F. Jackson and Thomas F. O'Donnell, eds. *Back Home in Oneida: Hermon Clarke and His Letters* (Syracuse: Syracuse University Press, 1965) .

# Chaffin's Bluff and the Darbytown Road

## September 28-November 4, 1864

*With the outbreak of the Civil War, the Confederate government began strengthening fortifications around Richmond. Stretching from the James River west of the city, the defensive works ran north and east of the capital, then turned south where they again anchored on the James. There, some eight and a half miles from the city, the engineers constructed Forts Gilmer and Harrison close to the Chaffin Farm and nearby Chaffin's Bluff. Located at a bend in the river, across from Drewry's Bluff on the opposite bank, this was a key defensive point on the route to Richmond. To assault the outer defenses, Gen. Grant ordered the Eighteenth Corps under Maj. Gen. Edward O. C. Ord to cross the James and move against the defenses on Chaffin's Bluff. The Tenth Corps, under the command of Maj. Gen. David B. Birney and reinforced by Brig. Gen. Charles J. Paine's 3rd Division of the Eighteenth Corps comprised of U.S. Colored Troops, was to cross a pontoon bridge at Deep Bottom and move toward Richmond via the New Market and Darbytown Roads. Led by Lt. Col. Francis X. Meyer of Utica, the 117th New York was about to encounter some very serious and deadly fighting at Chaffin's Farm and the Darbytown Road where diarist Fairhead came close to breathing his last.*[1]

---

Sept. 28th Wednesday. Warm. Orders for moving at 3 p.m. Were all ready in short order and then used the intervening time in writing. Active Rumor said that we were to take Transports. Wrote to my wife, also to my Mother. At 3 p.m. formed columns & moved off trudged along toward Point of Rocks, crossed on Pontoon over the Appomattox, passed Jones' Landing, followed down Jones Neck, & crossed a Pontoon over the James at that point. Passed through the fortifications on the North side of the James, stacked arms & laid down to sleep. This was at 2 a.m. of [Sept. 29].

Sept. 29th Thursday. At 4 a.m. we were aroused. The march had been long and wearying, many had fallen out. After eating something we marched within the enclosed works, & piled up our Knapsacks. Negro Troops has moved forward earlier than this. When the proper disposition of the advance was made, we moved forward as reserve. Skirmish-

---

1 Andrew A. Humphreys, *The Virginia Campaign of 1864 and 1865* (Edison, NJ: Castle Books, 2002), 284-85; Longacre, *Amateurs*, 211-12.

ing had commenced, and the Johnnies were throwing shot & shell among us. Some began to come to the rear wounded. There was a charge made in the front, & we moved farther forward, passing onward we came to the locality where the Negros had charged & taken a line of rifle pits. They suffered considerable here as a number laid dead, still we pushed forward. Skirmishers were out. Our Div. moved on another road & pushed rapidly forward reaching & passing a line of rifle pits, having Earth works for cannon commanding the roads, all skirting an excellent pine woods. Marching some distance into the woods, the Johnny shell began to screech through the forest. Our Brig formed line of battle across the road, some of our men were wounded & Killed by shell while this was going on. After forming we laid down a few moments. The Rebs had a battery of two guns a short distance in our front, which were sending canister among us. "Attention." "Fix bayonets." "Forward." "Charge." In an instant every man was up, bayonets on, & going forward. The cannon in our front, a large Reb batt. away on out "left," dealt death into our advancing column. Still on we went. Bullets began to cut down many. The ground was newly ploughed, men got exhausted & had to lay down & rest before they reached the position of the two guns. During this rest, the Johnnies "limbered up" & dusted.[1] Before this was done our men had breathed a little & were only up in time to give the "dusting" Graybacks a few shots. This position gained we came to a Halt. A church (or Temple Hall) stands on the height of the swell, and is on the turnpike. Along this pike a connected line of battle was formed. This pike thus parallel with a strong line of Forts, batteries & rifle pits, all crowning a ridge of hills ¾ of a mile distant. The intervening ground is very rolling, in the gullies there are swamps. The whole of this there is covered with slashed timber. Our Brig charged from the pike, over this, to near the strong line of works in the p.m. but were unable to carry them. The work of death & wounding was too fearful, the first column faced round, & fell back a little, but a small reserve of negroes coming up, a momentary rally was made, but this too gave way, men dropped with fearful rapidity while we were in a corn field just fronting the works. Our Brig lost full ⅓ of the number who started in this second charge. The more slightly wounded, & those who were wounded further from the Reb works were got away, but a horrid sight of Killed & wounded remained in the corn field. Our shattered ranks fell back to the road & formed in line again & laid down to get a little rest. The terrific cutting we had got, & the Extraordinary exertion we had been forced to make in climbing, tearing & crowding

---

[1] Used colloquially, "dusted" or "dusting" meant defeated or leaving/retreating.

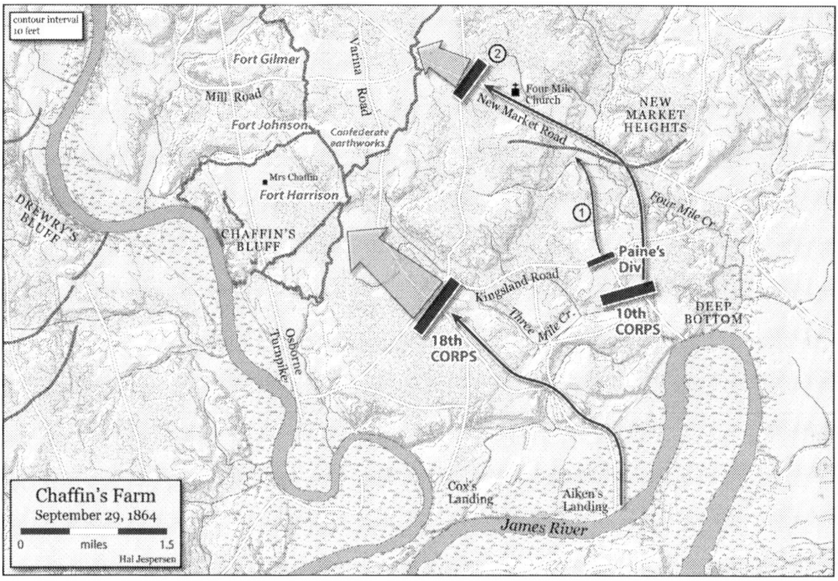


① Paine's U.S. Colored Troops capture the first Rebel rifle pits. ② The 10th Corps captures fieldworks at New Market Heights, turns left along the New Market Road, breaches the Confederate defenses and advances toward Fort Gilmer. *Hal Jespersen.*

---

over the slashing had wholly exhausted us. 3 swampy ravines were between the pike & forts. Reb batteries raked each ravine. While the 2nd Div. of the 10th A.C. was busy at this work, the 18th A.C. had been successful on our left in taking two or three forts, containing 15 guns. The ground leading to them was easy of travel, as I learned by personal examination afterward. At darked, out Div moved back through the woods, and occupied the Reb pits. Here we laid down & slept. Our Pickets were on the alert.

Sept. 30th Friday. Changed position a little along the Reb pits. Rainy somewhat. The Johnnies attempted to retake the works captured by the 18th Corps. The Rebs got severely punished for this rashness. Some work between our batteries and the Johnnies.

Oct. 1st Saturday. Rainy. We are engaged in turning the Reb pits,[1] con-

---

1 "Turning the Reb pits" meant modifying the fortifications so they could be used for defense in the opposite direction than that originally intended.

structing abbattis, slashing timber, throwing up batteries. We feel determined to hold our own. We feel like dealing a hard blow at them to pay them for using us so rashly. Got letter from wife.

Oct. 2nd Sunday. Rainy & chilly still. We are without tents & some without blankets. Still we are in lively spirits & watching closely. Our ardor is not diminished. Still strengthening our works.

Oct. 3rd Monday. Cloudy & rainy. Busy about as yesterday. Rations plenty [?]. They have issued them at night, which is some what disagreeable, especially when dark & wet. "D" Co.s' loss in the charges of the 29th ult. was 1 Killed, D. McNab. 6 wounded, Corpl. Morrison, Pvt [s]. Sayles, van Dresar, Baldwin, Downes, Goeble. Missing Powers & Hughes, probably wounded or Dead.[1] Capt. Magill came back to the regt to day, has been home 15 days, after Recruits. He got none, all had been sent to our regt. He reports all well at home. Our tents were brought today.

Oct. 4th Tuesday. Some sunshine, but thunder head clouds float lazily. Very quiet. Cleaning up & put our tents into shape behind the pits. Wrote to Hannah yesterday.

Oct. 5th Wednesday. Sun hot. Thunder heads. Quiet. Rebs are busy strengthening their works. They don't seem very venturesome. We feel Confident of Holding what we have gained. Gradually tightening around the Reb Capitol. Cant say when it will fall into our hands. Affairs remained quiet during the day. Our remodeling of the Johnny works is complete.

Oct. 6th Thursday. Hot at morn. Cloudy & breezy at M. Crimson clouds at p.m. Nothing of moment is transpiring thus far. Our men & the Johnnies have been exchanging papers. This was done yesterday, also Recruits are still coming forward. 3 came to our co. yesterday. Some of the regts are being paid each day. Some anxiety is manifested by some of our men for their pay. I am not eager for mine.

---

[1] David McNab, born in Scotland, lived in New York Mills. James Morrison, Utica plumber; Daniel Sayles, Utica farmer; Francis van Dresar, Westernville farmer; Henry W. Baldwin, Sangerfield shoemaker; Michael Downes, Utica farmer; Charles Goeble, Utica upholsterer; Sylvester R. Powers, Utica carder; Hiram H. Hughes, Remsen laborer. Powers was captured and exchanged near the end of the war. There is no further record for Hughes who was most likely killed in the attack.

Oct. 7th Friday. Warm. Waked at early morn. Engagement on the right. Cavalry men rushing in being much frightened. All the Troops were ordered to strike tents & ready for instant service. The Cavalry on the right flank seems to have been badly trapped. Changes in the infantry line were made. Our Brig was ordered from the Entrenched line, & marched to a position farther to the right, which was nearly at right angles with the Entrenchments. Formed line of battle & moved forward, heavy musketry in our front. The 1st Div formed a line of battle there, they were Engaging a column of Rebs who had crowded down the right flank. Our Brig cheered lustily as they moved onward, the line in front responded. Soon the firing ceased. They had come within easy range of our men, & had been severely repulsed, as they made no more advance. Our Pickets pushed forward. The Reb Killed & some of the worst wounded were left where they fell. Our men brought in the wounded. 40 or 50 Johnnies threw down their arms & rushed in. They evinced much pleasure in getting behind our lines. Once, a supple youth, jumped about considerable, dropping down behind stumps, to protect himself from supposed danger. His evolutions were laughable. He jumped up nimbly when he was ordered to assist in taking a wounded Johnny to the rear. He had an old hat on his head and a better one in his hand. Both lines remained quiet for an hour or so. Our Brig was then ordered to throw out a line of skirmishers and advance beyond the line formed by the 1st Div. Here we rested awhile, then moving forward, our Brig marched in Battle line through the woods & over an extensive clearing to works which had been occupied by our Cavalry pickets before the advance of the Rebs. The 1st Div. then came up and extended a battle line farther out to the right. We laid down at the old Reb earthwork. Had seen no rebs. The Johnnies had had a battery in position at the point where our regt came up in the line. They had their horses killed at that point by our arty. Thus the Reb endeavor to turn our flank was frustrated, & they left in haste, as they threw away considerable of their ragged Kit, Shovels, picks, & forage was left where their Battery was stationed. We laid down here & rested till 11 p.m. Then we moved to the rear of the entrenchments. Stacked arms, & made our beds.

Oct. 8th Saturday. Windy & Cold from the North west. Were roused at 4 a.m. Remained unchanged till 12 M. then moved a little farther to the left, & formed column in a forest, a little more to the rear of the entrenched line. Stacked arms & put up tents. A Picket was sent out at 5 p.m. Wrote to Hannah.

Oct. 9th Sunday. Chilly North W. Wind. Quiet here all night last. Remained in camp still on Reserve. The men have many fires, and sit around them most of the time, employing the time in the usual frivolous conversation of camp jargon. Largish whiskey rations were dealt out yesterday. A largish number of men were wildly hilarious during the evening, & were as usual full of rediculously, inconsistently mirthful. Some of the men have succeeded in getting more of the potion to day and are not a whit more sensible than last eve. A terrible curse to our army. Evening passed off quietly. Wrote to Mrs. Whetmore.

Oct. 10th Monday. Chilly as yesterday. Wrote to my Parents. Some firing from the Johnny Rams.[1] Not rapid. It is far to the left. We can not see it, only hear the reports. Had Co. Inspection this morning. Corpl Garlic came back to the Co. last evening. He received a very slight wound on the 15th of June at the charge on the Reb works before Petersburg, Va. Considerable white feather[2] among the series of actions from that time [?] forth to the present as regard his life.

Oct. 11th Tuesday. Dry & cool. In camp, had squad drill in the a.m. Co. Drill in p.m. Boys whiled time away about camp. Large fatigue party from Regt at work on fortifications. Ex Col. White[3] came as agent to receive votes. He makes a different appearance in citizens' than in Soldier garb.

Oct. 12th Wednesday. Cool in Morn. Cloudy at Noon. Rainy at Eve. Nothing in the drill line took place During the a.m. Boys at their usual work, cooking, mending, hunting lice (or in the soldier parlance Skirmishing). At 2 p.m. orders to get ready by 3 p.m. to move in light marching order. Were ready. Started. A largish column had gone ere Our Brig started. One Battery. Our Brig went near the outer fortifications, & halted in column, stacked arms, & rested. Rain began to fall. Toward evening the move was countermanded & we returned to camp, glad that we were to rest in our tents during that rain.

Oct. 13th Thursday. Chilly, brisk wind from the west. Sunny. At 2 a.m. all were roused. Packed our dudds in our knapsacks, reserving our rubber blankets for light marching order. Moved off at 4 a.m. having had

---

1 Ships fitted with a metal bow for ramming. They were firing from the river.

2 The term "white feather" was synonymous with cowardice. Fairhead is suggesting that he remained away from the regiment for a longer period of time than necessary for the "slight" wound he received.

3 Alvin White of Utica mustered in as lieutenant colonel and advanced to colonel. He was wounded at Drewry's Bluff and discharged for disability on July 18, 1864.

breakfast. Took position along an old reb rifle pit, situated to the right & outside of our fortified position. Another column extended from our right away to the Darbytown road. At Day light, the right began to swing forward, skirmishing commenced, & at times was very active, resembling a continued volley. This was the phase of the days work, our regt remained stationary only sending out additional pickets.

Demonstrating all day was the order. At dark all the troops had returned to former camps. Cant say what were the results, only that the Johnnies skirmish line was driven somewhat.

Oct. 14th Friday. A glowing Autumn day. In camp all day, drew rations, all quiet. Got letter from Hannah. Voting progressing finely, the large Majority going for the union candidates.

Oct. 15th Saturday. The gloaming[1] glorious. General Inspection at 9 a.m. Drill at 2 p.m. Dress Parade at 5:30 p.m. I cast my vote for the Union candidates.[2] Wrote to wife today. Gathered some dry brush and sticks, & had a cheerful bivouac fire in front of my tent during the evening. Would look at the glowing embers, & sparking blaze and contemplate the rusticity of this situation in contrast with the social association of Home. The wind wanders lively while cogitating on such topics.

Oct. 16th Sunday. Another glowing Autumn day. Company Inspection at 8 a.m. The fatigue Detail was called out to day. There seems to be haste in perfecting our works. Employed the a.m. after Inspection in singing, reading my testament, and <u>wrote</u> a <u>letter</u> to <u>Hannah</u>. During the p.m. was busy with drawing and issuing 2 days rations. Sent an old Edition (1814) of Esops Fables to my wife. Dress Parade at 5:30 p.m.

Oct. 17th Monday. Pay Rolls came during p.m. Some signed them. War quiet reigns. Commenced altering a pair of pants which I drew on Sunday. Worked about 2 hours upon them, much more to be done. Drew fresh bread. <u>Wrote</u> to <u>Charles Goeble</u>, sent him 50¢.

Oct. 21st Friday. Cool & windy. Recd orders to act as Sergt. Maj. In place of R. Bryan[3] promoted to 2nd Lieut. Mustered in "G" Co. Got

---

1 Twilight, dusk.

2 Absentee ballots began during the Civil War to allow soldiers to vote. This was the first presidential election in which they were used. Abraham Lincoln was the Union Party nominee.

3 Born in Scotland, Robert Bryan was a Kirkland bookkeeper.

letter from wife and immediately answered the same. Further good news from Sheridan.

Oct. 22nd Saturday. Lowery & cool with slight spray of rain. Wrote short letter enclosing $50.00 to Father. Co. Drills under command of officers, during a.m. 2 ½ hours. Battalion Drill in p.m. Some errors made in the handling of the regt. Maj. Bagg is not so familiar with the maneuvering of a regt as he is with the evolutions of a co.

[Oct. 23rd] Sunday. All quiet, coolish Fall day. No Drills. Dress Parade at 5 p.m. Spent some time reading the Word.

Oct. 24th Monday. Stiff winds. Short Battalion Drill at 8 a.m. to correct some errors of Saturday. The interval till 1 p.m. was expended in polishing straps and belts, and cleaning arms. At 1 p.m. regt formed on colors. Moved to the opening at 1:30. Formed Brigade line at 2 p.m. Brig. Gen. Foster[1] and staff came to review us. The Brig passed round but once at quick time. Review concluded we returned to camp by 3 p.m. An easy review. The whole affair passed creditably. Wrote to E. D. Williams.

Oct. 25th Tuesday. Invigorating weather. Wrote to wife sent $40.00. Commenced to rearrange camp, forming Company streets. Mr. Crocker from Utica was at Hd Qrs this morning, getting Democratic (so called) votes. A few, who had decided not to vote at all, were prevailed upon to swallow the whole grist unsifted. Upward of 18 in the regt cast votes.

Oct. 26th Wednesday. Cold. Boys continued during Early day to cut, hew, fit, and build up logs. At noon rumor of movements began to circulate. Orders came for three days cooked rations. Were all prepared. Boys laid down to sleep.

Oct. 27th Thursday. At 2 a.m. all were roused, and ordered to strike tents and pack knapsacks, leaving them in a pile. Got breakfast at 4 a.m. & fell in. Moved to the right of our position. Marched quickly forward, formed battle line and then marched forward till we reach a line of vacated reb works. Here we halted. During the day our Picket line & some reserves were maneuvered considerably. At 4:30 p.m. our Brig & part of one other were formed in three lines of Battle. In this order we

---

[1] Brig. Gen. Robert S. Forster at that time commanded the 2nd Division, Tenth Corps, to which the 117th New York was attached.

William Waud's sketch of the Battle of the Darbytown Road. *Library of Congress* .

advanced toward a reb work, shot and shell began to fly as usual. Onward we crowded, some were falling. After getting about ½ mile through the woods the column laid down and took a little rest. The command "forward" was again given. I raised up and repeated the word and was just urging the men up and forward when a bullet went "thug" through the side of my neck. It knocked me down. In a moment I was conscious of the locality of my injury. Finding the Carotid artery was not severed I was aware that I should not bleed to Death. I laid here a little while to allow the bullets to slack a little when I worked my way to the rear. Reaching the Doctor I had him examine the wound. He had no time to dress it. I tied my Kerchief around it and getting into an ambulance, was carried to the flying Hospl. Many horrid wounds were being dressed. Doctor Carpenter[1] Dressed mine and then assigned me to

[1] Henry W. Carpenter, a physician from Holland Patent, was a regimental surgeon in the 117th New York.

a bunk which was occupied in part by an adj[utan]t who had a bullet wound through the leg just above the Knee (flesh wound). He moaned considerable through the night.

Surgeon Henry Carpenter.
*US Army Heritage Education Center.*

Oct. 28th Friday. At early morn most of us were put on board a transport which left Deep Bottom about 3 p.m. reaching fort Monroe about 10 p.m.

Oct. 29th Saturday. At early morn were transferred to the New Camp Hampton, U.S. Hosp. Wrote to wife and mother. Wound was dressed. We had good beds, and good care. Many of the wounds are of the worst character.

Oct. 30th Sunday. Wrote to wife, Mother & John. Wound doing well. A slice of toast & coffee for breakfast. A slice of bread, potatoes, piece of meat and coffee for Dinner. A slice of toast, egg and chocolate for Supper. Met a number of Boys from our regt.

Oct. 31st Monday. Wrote to wife and mother. Hospital life about the same as yesterday.

November 1st. [Tuesday.] Lonesome somewhat. Hosp life is considerably Hum drum. Wrote Home and to wife. Furloughs have been made out for us to go home, all that can ride.

Nov. 2nd Wednesday. Furloughs talked about but not here. Wrote to wife, wrote to mother, also a letter to John. Wrote 5 short letters for friends.

Nov. 3rd Thursday. Rainy. The furloughs begin to return. All may get here through the day. Wrote Home also to wife. My wound still doing well. Much better than I expected. Hope it will continue to improve. Some men went on furlough.

Nov. 4th Friday. Strong wind. Wrote a short note to Hannah. Wrote a short note to Mother. Furloughs came in the a.m. near noon. We were ordered to Dr. White's office for Transportation. Reported, waited, got no Trans. Wind to[o] high for starting. Steamer loaded with sick and wounded. Ordered back to Quarters. Came back punning Each other about passing our time on furlough at our Hosp Quarters.

---

*The assaults on the Confederate fortifications southeast of Richmond cost the regiment 176 casualties, 124 at Chaffin's Farm and another 52 at the Darbytown Road. While Fairhead recuperated in the hospital, the regiment went into camp, preparing for what the men thought would be a lengthy stay over the winter. Dejected over their losses, they were nevertheless cheered that Petersburg and Richmond remained closely besieged, Atlanta had been captured, the Shenandoah Valley had been cleared of Confederates, and Sherman's army was marching seaward toward Savannah. It was clear the fortunes of war were turning decidedly in favor of the Union. The relative quiet of camp life in November gave the men a chance to relax and recover from the recent campaign, but always, in the backs of their minds was the nagging question "What's next?" For Fairhead, it was a continued "furlough" in the hospital.*

Chesapeake Hospital in Hampton, Virginia. *Library of Congress.*

Reunion ribbon from Fairhead's Company D. *Oneida County History Center.*

# Wilmington

## January 19-March 14, 1865

*While George Fairhead recovered from his wound, the 117th New York participated in two expeditions against Fort Fisher, North Carolina, covering the entrance to the Cape Fear River and Wilmington, the last remaining Atlantic Coast port open to Confederate blockade runners. The weapons and supplies they brought into the South were transported into Petersburg for the support of the defending Confederate army. Closing the port was essential. In the second attempt to take the crucial fort, the Oneida County regiment led the assault, breaching the Confederate defenses and being the first to place its colors on the parapets. This portion of Fairhead's diary begins with his hospital stay and the arrival there of casualties from Fort Fisher, then it continues with his return to the regiment and its march to Wilmington and beyond.*

---

Jan. 19th Tuesday. Fair. Wounded from Fort Fisher come in. Saw Gomar Evans[1] of "D" Co. was slightly wounded. Other Cos. had a number represented. These all came from the boat on foot & were dirty, not having had opportunity to wash since the severely & hand to hand contested fight. Went to Chesapeake. Found Capt. Magill with left leg amputated just above the Knee. Is in a fair way of recovery. Lt. Col. Meyer, a bullet through right fore arm. Skinner about the same. Bryan with slight wound on the left of the forhead. Others are wounded & coming. Saw Alfred Hull who left our company in July 1863 sick.[2] He had his hip injured when he got off the boat & is now lame. I finished Havelock[3] today. Wrote to Mother. Wrote to Wife. How sad that sin shall cause such suffering and sorrow. There is a nobleness in man to endure so much in combatting evil.

Jan. 20th Friday. Fair. Went to Chesapeake to day called on the Officers of our regt. They are full as easy as yesterday, rather more comfortable. Wrote letter for Lt. Col. Meyer. Wrote letter for Capt. Magill. Commenced to read "Catacombs of Rome." Called upon most of the men

---

[1] Gomer Evans was from Trenton.

[2] Lt. Col. Francis X. Meyer was a Utica salesman who led the attack on Fort Fisher. 1st Lt. Eugene C. Skinner was a student from Utica. 2nd Lt. Robert Bryan was a bookkeeper from Kirkland. Alfred F. Hull was a New York Mills farmer.

[3] A havelock is a cloth flap attached to the back of a military cap to cover the back of the neck.

who are in the different wards from our Regt.

Jan. 21st Saturday. Steady rain. Wind. Wrote to Col. Daggett 117th N.Y.V. Wrote to Wife and Father. Wrote these letters in Chaplain's tent of 2nd Division, Hampton Hospital. The tent is occupied & business dispatched by Robbins[1] of the 117th N.Y.V. and a Mr. Hart[2] of the 2nd Penn. Cavalry. Robbins came to the Hospl sick. Mr. Hart was wounded on left scapula: His arm remains weak. Called on our Officers wounded & at Chesapeake. All Doing well.

Lt. Col. Francis X. Meyer.
*Donald Wisnoski.*

Jan. 22nd Sunday. Rainy. Went to 2nd Division, changed my underclothes, blacked boots & put on a clean collar, then went to Church at the Bethesdean Chapel. Chaplain Marshall preached. Fair sermon. Wrote letter from Capt. Magill to his wife. Captain is improving, but is still sore.[3] The others are doing well. I saw Dr. Mowris[4] as I came from the Chesapeake. He is wounded on the shoulder from the Magazine Explosion after the capture of Fort Fisher. He is cheerful. Finished reading "Catacombs of Rome." Called on friend Robbins & spent evening with him.

Jan. 23rd Monday. Rainy. Commenced reading "History of War of

---

1 Probably a reference to Albert W. Robbins, an Augusta farmer, who had been wounded at Drewry's Bluff and was in the hospital at the time Fairhead wrote the entry.

2 There is no one named Hart or any close approximation of the name listed on the muster rolls of the 2nd Pennsylvania Cavalry. Since Fairhead calls him "Mr." rather than using a military rank, perhaps he was a civilian employee working with the regiment.

3 After leading the regiment into Fort Fisher as the senior officer present, Magill was hit by a shell fragment from a U.S. naval ship necessitating amputation of his leg.

4 James A. Mowris was a 117th New York regimental surgeon. He was injured when the magazine at Fort Fisher exploded after the fort fell into Union hands.

1812" by Brachenridge. War declared June 19, 1812. Called to see our Officers. They are doing well. Capt. Magill is much easier. Called at Ward 14 to see S. Inman,[1] he has gone home: started on the 22nd. Called to see Gomar Evans, he was not at his bed. Attended Prayer Meeting during evening.

Jan. 24th Tuesday. Clearing up. Called at the Chesapeake Hospl. Ate dinner in the Kitchen, had mashed Turnips, potatoes, meat, bread & butter with good coffee. Went to 2nd Division and stopped with Robbins during afternoon. Throwed chicken for John away, had mildewed. Sent paper containing linen collar to Hannah. Attended Prayer meeting in Dinning Hall.

Jan. 25th Wednesday. Pleasant, cold. Got Pass to Ft. Monroe, went through Fort. Got letter from Hannah, answered it. Wrote to Mother. Also sent paper & Key. Made some arrangement to go to regt. Called to see Officers. Some of them were in. Spent evening with friend Robbins.

Jan. 26th Thursday. Clear, cold. Rose at 6 a.m., got breakfast & started for Ft. Monroe. Waited all day about the wharf that I might learn something definite about a boat going. Learned that the steamer McClellan was to go to fort Fisher, got my Trans. Learned the steamer not going to day. Returned to Hospl., did not go out during evening. Got papers from Hannah. [???] & N. C. Advocate.

Jan. 27th Friday. Clear & cold. Arose at 6 a.m. got breakfast, two days rations and started for Ft. Monroe. Waited awhile for a chance to go on board the McClellan anchored out from the wharf. At last got on the Marthy Washington and from that to the Mc when the former came alongside the later. The Marthy Washington was busy all day bringing [???] to the McClellan. Cold & chilly. Had no blankets. About 9 p.m. found a roll of blankets not in use. Appropriated them for a bed for the night.

Jan. 28th Saturday. Cold. Mc[Clellan] started at 11 a.m. Passed Hatteras at night, rough. Sea sick. Slept but little.

Jan. 29th Sunday. Cold. Laid very quiet all day. My stomach not very easy. Became considerable settled by 6 p.m., reached the mouth of

---

[1] Samuel B. Inman was a painter from New York Mills who Fairhead considered "a ready, reliable soldier."

Cape Fear river, about this time, 31 hours from Fort Monroe. Fort Fisher and adjacent work loomed up like a ridge of successive hills. The McClellan anchored outside the Bar. Slept some what well. Weather considerably warmer than at Ft. Monroe.

Jan. 30th Monday. Fair. Debarked from McC about 9 a.m. Started in search of regt. As I advanced toward fort Fisher, shells & fragments laid as thick as coble stones in a creek bed. Hardly a step without touching one. Passed by the fort & found the regt. Two miles farther on. Glad to see me. I was pleased to [see] many of my old friends. Some are left but many away & gone. Got dinner of fish fried & hard tack. Regt got orders to take down tents, & be ready to march. Went to see Col. Daggett, gave him the checque sent from wife, by myself, learned the Regt was going to Battery Buchanan, for Q.M. services. Bat. Buchanan is near the wharf now building, & near where I landed, in a surf boat. The lumber used by the men in fixing Quarters is to be carried to the locality of the new camp. A team of six poor mules was furnished for "D" Co. I assisted John with his, accompanied the team, & had a high time in lifting at the wheels & pounding Mule meat. Teams got stuck. Procured two more animals, did some better. After a tedious two hours got a new location. Unloaded, & team went back. Fixed bed after eating supper of coffee & hard tack.

Jan. 31st Tuesday. Fair. Worked all day building new tent, at night had good shanty, fire place, bunk & sand floor.

Feb. 1st Wednesday. Fair & cloudy. Continued during day to work on Shanty. John & I walked about 1 ½ miles for purpose of finding a barrel to ornament and benefit our chimney which is made of sand soils.

Feb. 2nd Thursday. Clear & windy. Renewed duties of Sergt Major. Put a short fine handle into a new ax Mike Carry[1] found on board a steamer. Fried pork & soaked hard tack for dinner.

Feb. 3rd Friday. Lowery. Large details of men for fatigue.

Feb. 4th Saturday. Warm with clouds. <u>Wrote</u> <u>letter</u> <u>to</u> <u>wife</u>. Read 2 chapters in Bible relating to the adultery of David with Uriah's wife.

---

[1] Michael Cary was a Sangerfield farmer.

Feb. 5th Sunday. Warm & fair. Wrote to Father. Fresh fish for supper, fried in pork dripping. Passable coffee & hard tack. Got N. Y. Times first paper for a dime.

Feb. 6th Monday. Cloudy, brisk wind. Mail came to day, got letters from Mother. Sent a small piece of torpedo cable to Mr. Markham, was used by the Johnnies upon a complete torpedo system around fort Fisher, but which was rendered useless by our [use] of the Gun boat shells. Also sent a piece to wife, and piece to Mr. Ackroyd.

Feb. 7th Tuesday. Driving rain storm in a.m. Wrote letter to Hannah. Dry & warm in p.m. Recd letter from [Hannah]. Taliesan Evans[1] came to co. Dead Beating at Hosp. Others are on the boat outside the Bar. The rain dashed through our tent, very much, during the storm, this morn.

Feb. 8th Wednesday. Dry and windy. Aided in putting up tent for Adjt's Office. Wrote to Mr. A. Ackroyd. Aided in making tri monthly report. Not so busy as usual.

Feb. 9th Thursday. As yesterday.[2] Fatigue details not so many as usual. The 23rd Corps from the South West, com[mande]d by Schofield,[3] has been debarking during yesterday & to day. Ky. & Ten. Troops mainly. Ohio & Indiana.

Feb. 10th Friday. Clear & windy. Worked some on the Adj't's Office, am getting it into shape. Details have called out all the men. 1st Lt. Meyer[4] came back to day. Mail came, letter from Hannah & Annie. Wrote to Hannah.

Feb. 11th Saturday. Clear & windy. Every serviceable man on duty to day. Spent some little time in writing. A reconisance, at the front, re-

---

1 Tallison Evans was a student from Floyd who did not enlist in the regiment until January 9, 1864. Fairhead obviously believed he had been lingering in the hospital instead of reporting for duty.

2 Since this is positioned where he usually comments on the weather, it probably refers to it being the same as the previous day.

3 Maj Gen. John Schofield led the Twenty-third Corps that arrived as reinforcement to the troops around Wilmington. On arrival, he took command of the army as the senior officer present.

4 1st Lt. Spencer C. Myers, a butcher from Rome.

sulted in taking the Picket line of the rebs with some 60 Prisoners. No heavy fighting. The Monitor moved considerable close to the battery, up the river, a Double Ender followed somewhat nearly. They threw many shot.

1st Lt. Spencer C. Myers, Company G.
*Oneida County History Center.*

Feb. 12th Sunday. Clear & windy. Got out Guards, read some. Thru Kendall & myself went to the "Warrior" to get some old Lumber to fix Quarters. He went back on the "Genl. Howard" to send a small boat to take away the lumber (a nice lot). In half hour the small boat came only to tell me the regt had marching [orders]. I bid good by[e] to the Lumber & got in the small boat. Returning to regt found them all astir in packing for a march. Formed line at 4 p.m. and then broke ranks till orders to march came. Moved away at 6 p.m. A stiff wind blowing, so cold as to make us shiver to the marrow. Waited at the wharf 2 hours. Wind increasing to almost a gale. Men becoming numb with cold. Ordered to return to camp. Did so. Sand drifting like snow. Getting back, some put their shelters over their old frames. We soon had our tent roof on again, a fire in our sod fire place. We sat and toasted our shins, thankful the Expedition had not progressed on such a bitter night. Went to bed at 9:30 p.m.

Feb. 13th Monday. Clear & windy. Still in camp but preparing for going, giving out more cartridges, rations. Examining into the arms of the regt. The Johnnies are having big fires in the vicinity of fort Philipes or Anderson across the river. Wrote to James Morrison, Utica.

Feb. 14th Tuesday. Cloudy & windy. Still unsettled as to future moves. At 12 M. orders to march at 2 p.m. came. Moved to Brig Hd Qrs. Brig formed and moved forward about 3 miles, halted in a pine opening, got coffee & eat supper. At 6 p.m. fell in, took the beach and moved forward slowly. Pontoon Train came up. 8 mules to each wagon, and each doing its very best. Many of them became completely exhausted on account of the tremendous heavy wheeling on the soft beach. At 12 midnight reached extreme of the point made by the Sound which we were intending to pass by pontoons, a march from our camp of about 14 miles. Men exceedingly weary. The crossing was not made, a failure in pontoons or some failure in calculation. I judge the failure was in the miscalculation of pontoon mules. 12 to each wagon would have been nearer a just proportion of mule force. [As] it was there were as many whips and men to use them as there were mules. Such high pressure must exhaust the force long before 10 miles were traversed. At 12 midnight we began to retrace our steps, men already weary & angry. As soon as we got within our line of works men began to "fall out" by dozens. Getting to the place at which we took the beach we passed it by. Men thinking we were going to camp, getting about a mile further we were led back on another road through mud and mire. When we arrived at 4:30 a.m. to the locality we left at 6 p.m. of yesterday there were not a hundred men with the regt. 320 men went out with us. When we reached our stopping place it was raining briskly, a drizzle had been making travel very disagreeable. I picked up a pole, & fastening it to a couple of trees, soon had our tent up in some sort of shape to prevent the rain from venting its full force upon us. Spread our blanket and on the morning [of February 15] commenced a wearied sleep.

Feb. 15th Wednesday. John Sinclair, McNab, Michael[1] & myself laid half wet through and slept soundly till 10 a.m. Then waking up half rested, sore & stiff. Rain continued to fall till 11:30 a.m. then became warmer, but clouds in plenty. Men are coming in from their last night's "falling out." Got breakfast at 11 a.m. A failure in the attempt to "flank" the Johnnies by a pontoon across the Sound. Continued in same position all day, refixed our tent, making it more comfortable and tight. Camp very irregular. Made up bed at 7 p.m. Blankets half wet.

Feb. 16th Thursday. Warm and windy. Wrote letter to wife. No move as yet. Had a sound night's sleep though our blankets were not dry. Mended my haversack which became very much torn on the last march to the Sound. Some of the "stragglers" are coming back to day. Some of them are greatly fearful lest they will be cau't with the regt. should it be forced to fight. Got letter from wife. Wrote short letter to Mrs. Catherine Williams. Went to bed at 8 o'clock p.m.

---

[1] Henry "John" Sinclair and William McNab were both born in Scotland and worked as bleachers in New York Mills. Michael was probably William Michael, a Utica machinist in Company D with the others.

Feb. 17th Friday. Warm, clear & breezy. Arranged camp in column by Divisions. Got tent up in pretty good shape. Wrote to sister Annie. Straped razor for Eva Jones.[1] Kendall came to camp the first since out hard march of the night of the 14th inst. He went to our old camp. His back became lame, so much so as to over come his will. Some have not come yet. At 3 o'clock p.m. were ordered to be ready to move at dark, having three days rations and 60 rounds of ammunition. Boys packed Knapsacks. Drew 1 day's rations to make the compliment. Towards evening took down tents & packed them. Laid about or set around a bivouac fire and passed the time in punning. Dark came but no orders to move. About 6:45 p.m. the A. A. A. Genl.[2] of our Brig came to us & said it was likely and most probable we should not move. Capt. Downer[3] said tell the Orderlys to have the men put up their tents and make themselves comfortable. A general yell was rolled out by the men and in 15 minutes nearly every tent was again up and men settled for the night. Sergt. Shorey[4] was mustered as Second Lieut. to day. A very strong wind blowing at dark. Wrote to Hannah, made my bed and laid down to "sleep and pleasant dreams." My senses began to ??? in the borders of the land of rest when I heard some one calling for the Sergt. Maj. I answered. "Wanted at Hd Qrs." Went: "Tell first sergt. to get the men up, 'pack up' heavy marching orders: move immediately." Delivered orders; then packed Knap Sack. Regt formed line at 9:30 p.m. Moved forward. Our march was slow, frequent rests; reached Federal Point at 1:30 a.m.

Feb. 18th Saturday. Moon clear, with cool breeze. Our regt got onboard the "Eliza Hancox" a harbor tug, and were taken to Smithville, debarked, marched through the ville and went into Column by Divisions. This was 4 o'clock a.m. The men soon made beds, & as soon were sleeping soundly. I was oblivious to everything till 7 a.m. Waked up, quit my bed, had had a capital sleep, fried pork & hard tack, made coffee, then called John to help eat breakfast. Had a good wash, packed my Knapsack, then walked through Smithville. A large mound fort overlooks the wharf. The village is laid out some what regularly. The Street facing the sea or bay has a few quite pretentious houses. Streets run at right angles from this point. Fences are old with no paint. A brick court

---

1 Evan Jones, a Utica farmer.

2 Acting Assistant Adjutant General.

3 Because of all the casualties, Capt. Edward Downer of Utica was then commanding the regiment.

4 Eben M. Shorey, a Utica painter.

house, an old church, one or two largish buildings I presume formerly used for mechanical purposes, occupy a central portion of the ville. There are no fences about these last. The streets are generally broad. Many live oak trees are scattered through the ville, and were the houses blessed with architectural style, and a fair coat of paint, the ville would really be picturesque. It is capacious enough to contain 1000 inhabitants. Morning bright & warm. At 11:30 orders came to pack & be ready to move in half an hour. Moved at 2:30 p.m. Marched leisurely till 6:30 p.m. Halted in column by Div. Sent out Pickets. Built fires. Made supper. Went to our Downy [???] couch at 8 p.m. Had a good wash. An incident of the march, a soldier running with a bee hive on his shoulder, bees became too severe, he drops the hive, fights bees a little time, then grabs it again, finally is forced to leave it near the fence. One of our regt. drove the butt of his musket through the hive. A general rush of men for the honey, a general assault on the men by the bees, a large number of men got a handful of honey, the sweet morsel stringing through their fingers. A universal laugh from the men as they watch the scene. 4 negroes came into our line as we passed along the road. One Joe Jenkins is with us. He is a very sensible Dark, came from Fayetteville, N.C., about 100 miles up the river. Has been a week reaching our lines. He followed the sound of the cannon. He would not come into our lines till he could see our colors, as soon as that was seen he said to his comrades come we are all right. Joe is a young man.

Feb. 19th Sunday. Clear and warm. Slept soundly all night. All waked up at 4 a.m. and got breakfast, cooking it over our pitch pine fires. The smothering smoke of pitch gives us all the color of half breeds, and inflaming our eyes gives us anything but an appearance of beauty. The men are in fine spirits, as we loiter and hear the men pass the time in rigging each other with puns or acts. Our bivouac is in a pine opening. All trees large enough are cupped. Barkless, half charred trunks are more numerous than growing trees. These make rapidly burning fire wood. A large number of men are keeping up a chaw on pine gums. At 11 a.m. column moved forward meeting no opposition, reaching "Fort Anderson" at 5 p.m. The fort had been evacuated during the night, the Johnnies finding we were getting around their flank. The Fort is a very strong work, composed of immense mounds. 9 heavy sea coast guns were left. Expected to take boats & cross the river. Night came on. We built large fires, laid quiet till 4 a.m.

Feb. 20th Monday. Clear & windy. At this time [4 a.m.] we were waked. Marched to the dock got on board the Genl Howard steamer. Carried up & across the river, landed with Small boats & pontoons. Got into mud half way to the Knees in getting from the edge of the river to good ground. Made coffee & had breakfast. Formed column and moved forward. A slight amount of Skirmishing by the negroes in our front. Made an advance to near a line of reb works, formed one line of Battle. Disposed the balance of the troops in Column. Ordered to make ourselves comfortable for the night. Made fires & were soon at our ease. The Johnnies threw a few shell at the line of Battle. Did no Damage. Cooked Supper. Had coffee some pork & a few crumbs of hard tack.

Feb. 21st Tuesday. Clear & windy. Woke up at 6 a.m. A small allowance of hard tack was issued. Got breakfast. John's haversack was stolen in the night. It had 4 hard tack in which we had got for breakfast. We had 7 hard tack, coffee & pork for Breakfast. At 9:15 a.m. our regt was "fallen in" to cheer for the Capture of Charleston by Sherman. The cheer had gone the round of the troops, regt after regt. Fresh meat & hard tack were issued for the day. At 10 a.m. we moved toward the left. Demonstrated on the Enemies right all the p.m. Our regt was almost all deployed as Skirmishers and Flankers. At one point we got near a reb battery. A few in the column in advance were wounded. Capt. Granger of Genl Ames staff had a horse killed by a solid shot going through it; the Capt. escaped with no farther injury than a bruised leg. Skirmish firing was Kept up at this point during the balance of the p.m. We did not remain here but went farther to the left until we rested on the river. The fleet was well up on our flank and sent in a heavy fire during the 2 hour following 4 p.m. Fortifications were being thrown up. Work was ordered to cease at 12 Midnight.

Feb. 22nd Wednesday. After ceasing work, we sat around till a little after one a.m. When we began a retrograde movement marched till we got to the locality of our last night's rest, reaching it at 2 a.m. Here the Orders to draw rations & be ready to march at daylight. On account of the drawing rations many of the men did not get any rest during the night. John & I made our bed & laid down. The locality where our regt were Bivouaced was a low, mucky & damp piece of ground which having recently been burned over by the grasses catching fire was black [as] tar. John drew the hard tack & we laid down & got 3 hours sleep. Up at Daylight, made coffee, fried pork & "hard tack." Ate breakfast. While enjoying this repast, word came to regt head Qrs that Wilmington was evacuated. Soon after we moved from our Bivouac, going in the direction of Wilmington. Passed through two lines of Fortifications. The Second & nearest to Wilmington was strong & difficult of approach being fronted by a continued system of ponds made by throwing causeways & dams across the low & narrow place through which the water would naturally have made its way. These ponds would make a very unpassable obstruction to any advancing column. Heavy guns were so planted as to rake the causeways leading to the fortifications. At 10 a.m. I first caught sight of the white Flag flying over the court house of the city. We halted outside about an hour & ¾. Then moving forward I passed the first house in the suburbs at 12 M. The large majesty of the houses are low one story board buildings, many of them having a rude piazza to suit. Very few of them has even a coat of white wash. There are some fine brick residences, 8 or 10 churches, some of very pretentious architecture. Two very nice ones which were built as far as they now are, were commenced in 1859. Slate roofs, lowering steeples, and Gothic style of brick. The uncovered points of the structures are weather worn, the rough boards nailed in the windows have the color always, marking timber exposed to weather. Capital to be invested in church property must have very suddenly dried up. The streets are sandy & soft, one short strip of sidewalk was brick pavement, but it had been buried by about two inches of sand. A large number of children & women showed themselves at windows, on corners, on Balconies, over fences. Old negroes were ready with their blessings. At one house a fine American flag was hung out. One fine looking young Lady stood on a corner, waving a small Union flag as we passed. I gave her a salute as I passed. Most of those who gazed at us as we marched through appeared as though they had recently combed their hair as it was very sleek & appeared wet. There were a largish number of white men about. They all say that the only way they could remain was to hide, as the reb Troops were leaving. We marched through the Town, passing through the inner line of Fortifications, over a bridge about ¼ of a mile long. As we were moving toward this bridge two men came running into our line, on the left flank, calling out "How are you Yanks." One was a reb Deserter, the other was an escaped Union Prisoner. He belonged to the 5th Iowa Cav. [and] was taken prisoner at the time Genl Stoneman was when he made his raid toward Macon, Georgia. He was thin. After crossing the bridge we "filed" to the left. Went into column by regts. Bivouaced. Some put up tents. A House was near our regt. This was taken as for Hd. Qrs. I taking one room for Adj't's office. Went to bed at 9 p.m. Washington's birth Day.

Feb. 23rd Thursday. Cloudy. Woke at 7:30 a.m. The rebs had rifle pits running through the Streets of Wilmington. They did not fight as desperately as they intended when they dug them. They evidently have been forced to change their minds. "Bully" for the strength of Uncle Sam's army. Wrote to Hannah. A very fine Cemetery occupies a section of ground on the North West suburbs of Wilmington. A heavy piece of cannon is in position near a section of the Cemetery which has the evidences of having been used for burial of the poorer classes. The distinction made by man are not those which actuate the judgement of Him who gives the final sentence of all men. Distinctions made by men will then in many cases be reversed. The rich will be found beggarly poor & those by man's understanding called poor will then have treasures of far more value than rubies. At 2:40 p.m. I got a piece of fresh beef which had been foraged by "I" Co. I soon had it over the fire boiling for supper. Some of the men came in with carcass of Sheep. The stock about here is lean. The good having been appropriated or Driven off by the Johnnies. The soil here is composed wholly of sand. Decomposed vegetable matter gives some of it a darkish color. The country just around Wilmington is somewhat cleared & had been cultivated, but out 3 miles there seems to have been no cultivations at all. Scattering pine forests, every tree large enough endure it, has been cut & cupped for purposes of making turpentine. Many of these trees have been on fire and the cut sides of the trees are as black as coal, many of the trees have been Killed by the fire and now the stark, barkless limbs and trunks, still standing, give a dreary desolation to the whole scenery. The whole country has the appearance of having been forsaken by humanity altogether. In marching from Smithville to Wilmington 30 miles over roads no better than a cow path we met but two or three board shanties. A house was out of the order of things. A wolf would starve in this country. At 5 minutes of 5 o'clock p.m. 2 men of "E" Co. came to camp having 2 fore Quarters of young beef swinging on a stick between them. Sound on confiscation. Our fresh beef is boiled, coffee making for supper, soon we will have a gay meal. Wrote to Fred E. Garrett.

Feb. 24th Friday. Rain all day. Went into Wilmington, Down on the river side. The streets & walks are paved. There are many fine buildings in that section. There are also Some capital public buildings. More of a city appearance in that section. Sign boards are common. Got considerably wet on this tramp. Wrote letter to Mr. Markham. Wrote to Gregory at Hampton Hospl. Rain ceased at 7 p.m. but cloudy. Went to rest at 9 p.m. Made bed on the floor, hard but better than the wet ground.

Feb. 25th Saturday. Rainy. Guard mounted at 9 a.m. Picket sent out some what later. Mail came, bro't letter from wife. A Treat. Wrote to J. A. Quinn, also to Ed E. Tubbs Mother concerning Ed E. Tubbs who was taken Prisoner Aug 25/64. John & Kendall went to the city this morning, bought 10 lbs meal, 5 lbs Sugar. I made a pot of Indian mush as soon as they came back, and had Sugar & mush with good coffee for dinner. They bro't back about 3 Qts of Peanuts. I roasted them in a spider,[1] we ate them while warm, they were excellent, very much more agreeable to the taste than when cold. Hot Peanuts are the thing hereafter.

Commenced to read "Life of Walter Scott." The book was appropriated by the Ad'j't from a vacated residence. A splendid beauro, looking glass, a beautiful painting [with] gilt frame and a number of other articles to make himself comfortable while remaining here, were appropriated at the same time. Some beding & chairs with tables were brought away. Soldiers who have left homes must be made comfortable by articles in houses left unoccupied.

Feb. 26th Sunday. Sultry with showers. Had regular guard mounting at 8 a.m. Got pass, went to city. Attended church. About 300 Soldiers present. A number of citizens were also there. A young man preached, stationed pastor Text Luke, 23-46. He said not a word regarding our Nation either in his prayer or Discourse. The church was a substantial brick edifice finished to the ridge in side Oak stained wood work. I walked about the city a little. Saw a few very artistically arranged gardens. Returned to camp. Wrote to wife. Baggage came this p.m. Our Chaplain gave us a short sermon in the p.m. Conklin,[2] Green, Kendall and myself did the singing. We also had an agreeable sing during the evening. Our Books were in the regimental desk. Wrote another letter to Hannah.

Feb. 27th Monday. Cloudy. Busy all day writing, making our Rolls and copying orders. Order promoting me to Sergt Maj. dating to Jan. 1st 1865 was made out today. Wrote letter and sent package to Mrs. Edward Griner whose Husband was buried by the Explosion after the capture of Fort Fisher.[3] Wrote letter to Mother. John got letters from Home to day. Had some more mush and sugar for dinner. This is a luxury far above the usual order.

Feb. 28th Tuesday. Cloudy. Busy all day making out papers pertaining to muster rolls, and returns. Large Details for picket were made to day. Larger hereafter than formerly. Post Guard is discontinued for the present. Commenced letter to wife, shall finish tomorrow if all be well.

---

1 A reference to a cast-iron frying pan with legs made to stand among coals for cooking.

2 This may have been a reference to Thomas Burgess Conklin, a farmer from Western in Company I.

3 Edward Griner, a glassblower from Verona.

March 1st Wednesday. Cloudy with rain. Finished letter to my wife. Sent filled Diary Home this morning; have been busy all morning. During evening read in Memoir of Walter Scott. Born 1771. Admitted to the Bar 1791. Married to Lady of French decent, 1796.

March 2nd Thursday. Cloudy. Monthly inspection at 10 o'clock a.m. by Lt. Nixon of Daggett's Staff, A. A. Insp. Genl. Had beans for dinner of my own boiling, they were very good. Gave Chaplain some. Sang some early in the eve. Got letter from wife, also one from cousin Kate Beaufort, S.C. Wrote to Cousin Kate.

March 3rd Friday. Heavy fog at 8 a.m. Warm. Wrote letter to Hannah. Busy all p.m. in office. Sung some at evening. Read a portion on Scott.

March 4th Saturday. Cloudy. Wrote to Hannah. Got my warrant for sergt. Maj.s' position to day & did it up in shape to send Home. Read some. Wrote to S. Barrows in evening. Sung considerably.

March 5th Sunday. Cloudy and coolish. More quietude has characterized this day, than has marked the preceding ones since we have made our home here. Chap[lain] Jones preached a fair Sermon at 2 p.m. Have spent considerable time reading "Scott." No excitement. Sung some little with Kendall just before retiring which we did at 10 o'clock p.m.

March 6th Monday. Pleasant. In office all day wrote some & perused "Scott" to some extent. Men anxious for mail.

March 7th Tuesday. Pleasant. Mail came bringing me a letter from Mother, also one from my wife. Read some in Scott. Felt something of restiveness. Not much business in the off[ice]. Sang a little at evening.

March 8th Wednesday. Cloudy, few drops rain. Wrote letter to Hannah. Also to Mother. Rearranged camp to day. Laying it out at the west of the house and closer than it was before. This will be much better than before. Glad I am so Quartered as not to be forced to the work of change. Not much business in the office. Finished "Sir Walter Scott."

March 9th Thursday. Cloudy. Furious rain storm swept past during last night, waking me twice. Rain commenced to fall this eve very abruptly, coming down plentifully. The chaplain took up Quarters in his tent to day. Wind & rain is increasing now, 9:30 p.m. I have a sound roof over my head, but many have nothing but cloth. I recd $2.05 from Adjt. C.H.

Left, Adjutant Charles H. Roys (*Donald Wisnoski*); right, Acting Assistant Inspector General George Washington Ross (*US Army Heritage Education Center*).

Roys[1] for a p[ai]r of booties [?] charged to me, aiding him in making out his clothing Return. No mail to day. Wrote to John Ross, Utica. Sang an hour at evening.

March 10th [Friday.] Fair and stormy. Read considerably. Was in office most of the day. The camp was thoroughly Policed prefatory to Inspection tomorrow. Wrote to Hannah. Went to rest at 10 p.m.

March 11th Saturday. Read some. Monthly Inspection was made by A. A. I. Genl. George Washington Roys,[2] 1st lt. 117th N.Y.V. Commenced to peruse "The Two Rivals," a sort of Sentimental Romance, blending Burr & Hamilton. The representations were greatly in favor of Burr & the other extreme regarding Hamilton, hence I put the book aside.

March 12th Sunday. Invigorating. Walked with Sergt. Green to an old Reb camp. Service at 2 p.m. A Stranger gave us a short sermon. He was just from New York State. Mail came. Got no letters. Felt some what disappointed. But will wait patiently. Sang some during evening.

---

[1] Charles H. Roys of Rome was breveted major for gallantry at Fort Fisher.

[2] Actually, George Washington Ross, a merchant from Utica.

March 13th Monday. Superb. Made table and moved our bunk and traps down into the Room occupied as an office. Picket detail was lessened to day to 43 instead [of] 60. Order issued to have 3 hours drill each day, by companies. Had a very fine sing during evening.

March 14th Tuesday. Breezy, some clouds. Strolled over to & through the cemetery. It is a very beautifully ornamented spot. Neat trim, luxuriant hedges, Iron palings are common in connection with hedges. Many burial spots are terraced. Shade trees in plenty. Sinuous roads and walks. Flowers & ornamental shrubbery in abundance. Highly finished & artistic shafts mark the spots where sleeping mortals rest till the trump shall wake them from their deep slumber. The hedge rows & much of the shrubbery is of a deep green & covered with prolific foliage. A beautiful sequestered spot. At one extreme of the cemetery, there were a half dozen Soldiers busy digging & covering graves in which soldiers rest, freed from the toils of war. Many of them were the remains of Paroled Prisoners who through the hellish inhumanity of the Rebels had but life enough to be brought with in our lines, & that only to die. Many a head board stands at the ends of these graves with no other inscription than "Unknown." Where are the hearts that writhe in the crucible of uncertainty for the one there laid to rest and marked unknown. The shrubbery here is far superior to that found at the North. The very moderate weather of this latitude giving the superior capabilities for the development of vegetation. Orders to be ready for marching at 1 o'clock p.m. of tomorrow. One wagon, only, allowed for transportation, & that wagon to carry 8 days forage for the horses of the regt. Was busy all the after part of the day packing such articles as could be carried. All the surplus articles to be packed and Stored. Men are allowed 1 blanket or overcoat, 1 Rubber, their shelter, one change of under clothes.

---

*On the same day Fairhead penned the foregoing entry, Gen. John Schofield received an order from Gen. William Tecumseh Sherman directing him to march toward Goldsboro. At 1:00 pm on the following day the men formed ranks to begin what would be their last campaign of the war.*

# The End at Last

## March 15-June 5, 1865

*From Wilmington the regiment marched northward through Faison's Depot on the Weldon Railroad, followed by Kenansville and Mount Olive. After fording Allen's Creek, foraging liberally along the way, they linked up on March 21 with the advance elements of Sherman's army at Cox's Bridge on the Neuse River. Continuing on toward Raleigh, on April 10 a courier brought the joyous news that Lee had surrendered the Army of Northern Virginia to Grant.*

*The 117th New York continued its trek toward Raleigh where it awaited what everyone believed would be a final showdown with the remaining Confederate army under Gen. Joseph E. Johnston. Fortunately, before that took place Sherman met Johnston at a farm owned by James and Nancy Bennett near Hillsborough on April 26 where Johnston finally surrendered all remaining Confederate forces in Florida, Georgia, North Carolina, and South Carolina, more than 89,000 men. For all practical purposes, the long catastrophe of Civil War was finally over. Thoughts now turned toward home.*

---

March 15th Wednesday. Cloudy with some rain. At one o'clock, p.m. having been busy in getting ready, we were "fell in," and moved forward. Made about 9 miles & at 6 p.m. formed column & put up tents. At 8 p.m. laid down to rest. Rain fell rapidly during night. Our tents served us nicely.

March 16th Thursday. Breezy, hazy clouds. At 5 a.m. recd orders to be ready for tramp at 6:30. Promulgated the order. Then began to get breakfast. At the time specified, all were ready with knapsacks strapped upon their backs. Moved forward steadily till 1 p.m. Halted & had an hour for dinner. The soil is becoming clayey. Vegetation is some more luxurious than around Wilmington. Country unvaried in its general outline. The regt. has marched capitally today, making little or no delay when passing the low wet sections in the road, but splash through. Wherever these wet spots occur, the bottom is usually hard sand, held firm by the wetness. At 2 p.m. again moved forward. About 6 p.m. reach a ford about hip deep. Many took off Pants, drawers, stockings & shoes. Others left their drawers on, changing them for dry after crossing. I did the latter.

Went into camp for the night a short distance after passing the ford. Established Guard & picket & went to rest for the night with orders for revelie at 4:30 & march at 6 [a.m.].

March 17th Friday. Fair weather. Revelie & marching transpired as ordered together with the requisite amount of cooking. Moved forward briskly. The rain of last night had made the roads very wet. Many wet place to cross one as deep as the Knees. This wetness is bad for feet. Some places in the road is very muddy. The lay of the land is becoming more varied. Better soil, & more hard timber. Some of the men are coming in with geese, hams, Shoulders, bacon, chickens, mules, horses. Came to a halt at 11 a.m. Pontoons & trains are behind. Put up tents. Sent out foraging parties. I got a lot of pea nuts. John went out & got a fine shoulder & bushel of sweet potatoes. Had good supper. Foraging party came back bringing or driving in cattle & sheep. Went to rest at 8 p.m.

March 18th Saturday. Beautiful air. No haste in getting out this morning. Revellie at 6 a.m. Made breakfast of sweet potatoes, fried ham & tea. Ate heartily. Orders to be ready to march. Packed & sat about. Men having sport in one way or other. Packed up haversack full of shoulder, John & Kendall carrying hard tack, a large am[oun]t of beef, & pork is thrown about upon the ground to be left to rot. War is distructive to civil law and private property. Moved about 10 miles and then went into camp for the night under the protection of a pine forest. Land still improving very much.

March 19th Sunday. Pleasant. Started forward at 6 a.m. Soon after starting we were forced to ford Allen's Creek, about buttock deep. Men yelled lustily while passing through. Many of the men had breeches, drawers, & shoes & stockings off, their accoutrements & haversacks hanging on their guns. A fine spectacle to those who might have a fondness for the grotesque. Nude legs & buttocks were numerous. After passing this stream we had time to don our proper garb & arrange our accoutrements & traps. Being ready we moved forward. Our Brig[ade] Guarding the train. Our march was some what irregular, with some very long hitches. The country through which we have passed to day is fine. Undulating, and much improved. Ploded onward, very weary and foot sore, aching very acutely. Long hitches as we came near the place selected for camp. The front of the column got to camp by sun down. At 8 p.m. filed into a newly ploughed field, Regt formed lines across, stacked arms, & fixed for the night. Many were so weary that they made bed without supper. I & comrades made sup[per] and went to bed at 9 p.m. Kind enough for our bones. The 112th Regt which was in rear of the Train, did not get into camp till 2 a.m. of Monday.

March 20th Monday. Beautiful. Revellie at 5 a.m. Up, got breakfast.

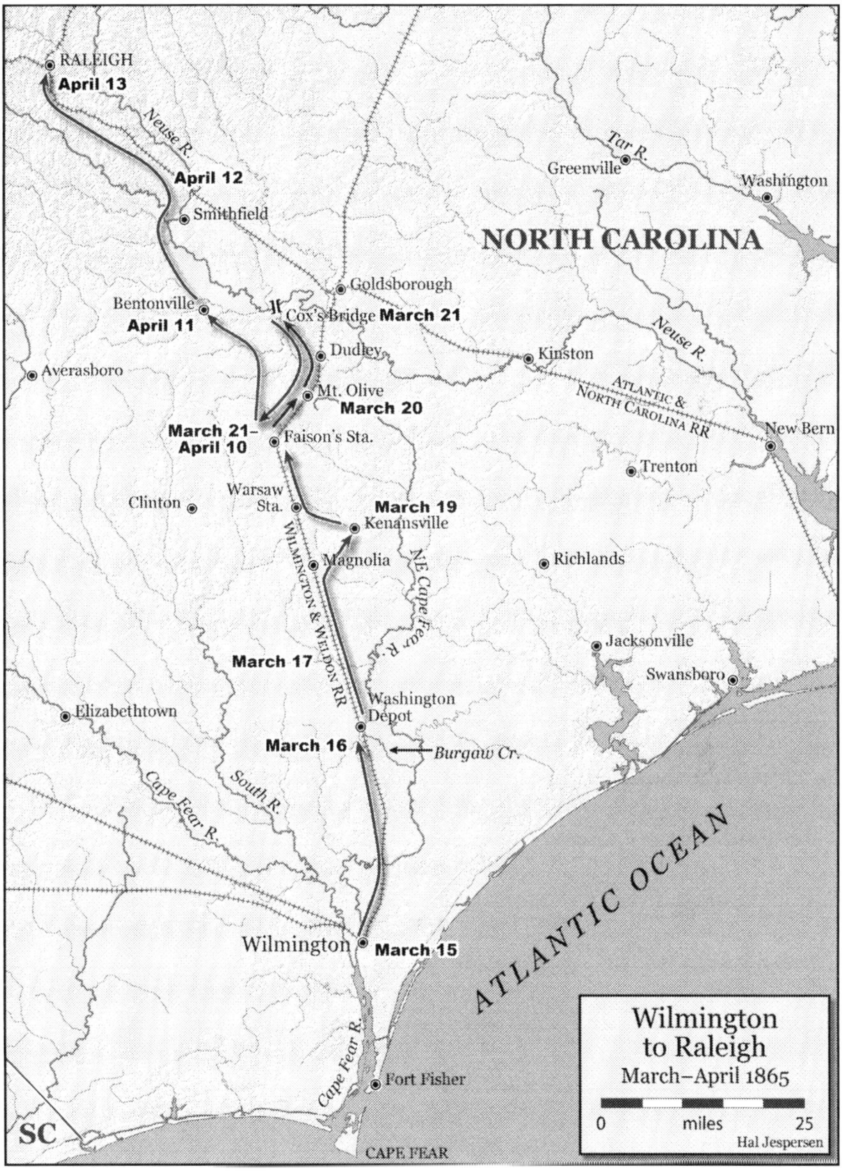
RALEIGH
April 13
Neuse R.
April 12
Smithfield
Tar R.
Greenville
Washington
NORTH CAROLINA
Goldsborough
Bentonville
April 11
Cox's Bridge
March 21
Dudley
Kinston
Neuse R.
Averasboro
Mt. Olive
March 20
ATLANTIC & NORTH CAROLINA RR
March 21–April 10
Faison's Sta.
New Bern
Trenton
Warsaw Sta.
Clinton
March 19
Kenansville
Magnolia
NE Cape Fear R.
Richlands
WILMINGTON & WELDON RR
Jacksonville
March 17
Swansboro
Washington Depot
Elizabethtown
March 16
Burgaw Cr.
South R.
Cape Fear R.
ATLANTIC OCEAN
Wilmington
March 15
Cape Fear R.
Fort Fisher
SC
CAPE FEAR
Wilmington to Raleigh
March–April 1865
0 miles 25
Hal Jespersen

Column moved forward at 6:45. Our Brig was 2nd in the line. Made excellent time, long hitches. One stretch was 67 minutes long, about 3 miles. We have had to ford a stream about knee deep. I passed it on a log way, made by morticing hewn logs upon firm upright stansions or posts. Still onward. Past through some good tillable lands. Heard of Sherman's Cavalry being about this section the day before. Toward night saw a few of the 17th Corps belonging to Sherman's Army. Men are very tired. Made camp at 6 p.m. in a fallow field. Cooked supper. Made out Tri Mo [tri-monthly] Report and went to bed.

March 21st Tuesday. Cloudy with some rain. Revellie at 5 a.m. The work of Breakfast & washing over, Packed up, ready for moving. Some little delay, not getting under way till 7:15 a.m. Moved onward steadily. Came across Foragers from Shermans Army, passed a camp or so that had been occupied by some of his army. Broken carts, carriages, dead horses, cattle & many other works of destruction were strewn rudely in confusion. Sherman foraging parties are a rough set. Many of them dressed partly in "reb" or citizen's Suits; some wholly. All are mounted either on mule or horse. Carts loaded, drawn by ox, mule or horse. Carriages & fine teams used to carry hams, shoulders or bacon, fresh meat, corn or fodder. Everything that can aid in giving support or sustenance to an army is appropriate[d] to this army or destroyed. Nearly every horse man has a ham or shoulder dangling from each side of his saddle, frequently a goose, pair of chickens, a brace of ducks. Our men frequently come into the line with from 1 to 6 hams strung on their guns or sticks, dangling over their shoulder. Passed on till about 1 p.m. when we halted, made camp, rain commenced to fall rapidly. I had my tent up in short notice. Sherman's foragers are scouring the country in every direction. Always going in parties or squads. Hams, shoulders, bacon, geese, chickens, ducks, corn, meal flour; every thing that is found eatable. Went to bed early. Have heard firing off to our left. Some of Sherman's troops are engaged, but have the best of the fight. Rain during evening.

March 22nd Wednesday. Bright, strong wind. Rain fell considerable during the night. Revellie at 6 a.m. No orders for marching. The Negro troop moved away at 11 a.m. Foraging & scouting parties are passing frequently. Our boys have been about some, had sweet potatoes & fresh pork for dinner. A foraging party of a hundred men was sent from the Brig in the p.m. The Q. M. wagons went along with them. Hope they will bring much meat. General Sherman called at Terry's Hd Qrs this p.m. I did not see him. In the fight of yesterday the 17th Corps had a severe part. The Johnnies were whipped, under cover of the rain &

darkness they sloped for the present.

Foraging party came back bringing meat, beans, molasses & lard. Kendall got 2 hams. Had supper of fresh meat, fried and boiled sweet potatoes. Ate heartily and made my bed and went to rest. Wind ceased at sun set.

March 23rd Thursday. Terrific wind, dust in clouds. Saw Genl Sherman at 7 a.m. as he passed by at the head of this Staff. I had no fair view of his features, whiskers & mustache trimmed close. He was smoking a cigar. The 14th & 20th Corps passed toward Goldsboro taking a road to our left. The 17th Corps passed on the road that runs by our front, going to the same place: all will cross the Neuse river. They made considerable sport at our camp guard & well they might. They are a rough set. Many have reb or citizen pants. Many have their coats. Negro Pioneer corps. Negro women follows the Trains, some on horses or mules. I saw a few white women following. A Foraging party was sent out to day. Kendall went again. They were attacked by reb cavalry. They dispersed & ran into the woods. Most have come back. Kendall does not show himself as yet.

March 24th Friday. Some wind. Moved tent into woods across the road. At evening, orders for marching at early morn were promulgated. Kendall & others came back after dark. They had seen no reb. Those who met the Johnnies, had a tall time in running to Camp. A man by name Stowe,[1] of "C" Co. is missing, supposed to be dead or prisoner. Kendall brought in a chicken & some honey, also sweet potatoes. These articles have been served up in due order to day.

March 25th Saturday. Cool & cloudy. Moved from our camp at 6 a.m. Marched to Fasion's Station on the R. R. from Wilmington. 18 miles or a little more was done by 6 p.m. when we formed camp in an open field beside the R. R. The country through which we have marched is very much under cultivation, and is considerably rolling. No board fences, except in front of three residences. The soil about this station is mostly under cultivation & is mostly level. A few low places through which small steams wind their way.

March 26th Sunday. Fine. Rearranged camp in accordance with regulations as far as possible with our limited number of tents. Company Streets about 16 feet wide. Officers tents near 80 feet in rear of Co.

---

[1] This may have been James S. Stone, a Utica shoemaker, who soon reappeared.

Staff Off. 40 in rear of Officers'. I & Kendall occupy the fly which serves as office. Men forage much. Orders against indiscriminate Foraging have been issued. Have Brigade guard mounted each morning.

March 27th Monday. Fair. Wrote to Hannah. Staff officers, & line off [icers'] tents not in right locality. Ordered to move them farther to rear, only a few feet was necessary. Kendall & I soon had our fly moved. There is much leveling of the corn ridges about and through our camp. "Airs"[1] are the order of the day.

March 28th Tuesday. Cool & cloudy. Brig Guard mount at 9 a.m. Adjt Roys officiating. Mail came at 11 o'clock a.m. bringing letters from mother and sisters, but none from My Dear one. Recruits or Drafted men are coming to the Division in large numbers. One large squad has already gone to the 142nd N.Y.V. A large squad is at Brig Hd Qrs now 3 p.m. Muster rolls of 200 men came from Harts Island, assigning that number to our regt. No recruits or additional men for our regt have made their appearance. Capt. Downer was some hilarious tonight, has been at Brig Hd Qrs.

March 29th Wednesday. Fair & cloudy. The system of "Calls" has been followed to day. Our dinner today was an excellent chicken stew. The material was the result of Kendall's ramble on yesterday: have one Biddie[2] in reserve for tomorrow. Such things are a treat for a Soldier. Two mails arrived to day bringing me 4 letters (two from Hannah). Dress Parade at sunset. Capt. Downer nearly able to hold himself in soldierly attitude.

March 30th Thursday. Strong winds, clouds & rain. Rain prevented the Drills. A considerable disagreeable day. Mail came twice bringing me a letter each time from Hannah. I also got one from John Ross. One from Mrs. Griner. Wrote in the office all the evening, & failed to write to Hannah. Was selected for recommendation for 2nd Lt's. Commission. At first declined, but on talking with Col. Daggett, and getting certain promises and statements from him, reconsidered my first decision and accepted the offer.

March 31st Friday. Windy, clouds and sunshine. Went at work making improvements on our fly. Made very decide[d] advance. Made two

---

1 Music.

2 A hen or small chicken.

good stools, have a fine bunk and can now enjoy our stay here very much more than before. Camp life has moved on in regular order. Inspections of Camp and Garrison Equipage have been made in several Co's. Much of the stuff has received the Excommunion Sign "II. C." and are taking from the amount for which some one is responsible.[1] Cooked another chicken stew to day. It was excellent. Wrote to Hannah.

April 1st Saturday. Sultry & cloudy. Fixed our chanty in better shape had bunk made. Usual camp duties were attended. Washed myself thoroughly in the brook running under the R. R. Some April fooling has transpired. Played little or no attention to it.

April 2nd Sunday. Warm. Company Inspections. J. D. Jones preached a short sermon at 2 p.m. These sermons have a very perceptable effect in quieting the turbulence of many men. 125 Recruits came to the Regt to day, and were assigned to the companies of the Regt. A large majority of them are foreigners.

April 3rd Monday. Cloudy with rain. Recruits are being got into order for service. Mail brot a letter from Hannah, which I ans[wer]. Had a sing in the evening. Some of our Recruits are good singers and have been singing a number of sentimental Song very nicely. They while the time away very cheerfully.

April 4th Tuesday. Warm, some clouds & wind. Usual camp duties. More attention paid to drilling by the officers. Had Battalion Drill in the p.m. Wrote to Father. Also sent a letter to Brother James. The Recruits are drilled with the arms of the veterans, a few of them are soldiers.

April 5th Wednesday. Sun hot at 9 a.m. Cloudy with wind. A few drops of rain. Windy at eve. Battalion Drill. Had a fine sing at evening. A fat ox which a foraging party brought in yesterday was missed this morn. Some person must have led it way during the night. Capt. Erwin[2] came to the Regt, has been absent since Aug 25/64 when he was wounded in the arm.

April 6th Thursday. Floating clouds, warm sun. Usual drills. Worked most of the day in fixing over a p[ai]r of new pants. Battalion Drill in

---

[1] What he means is that some unauthorized units were drawing supplies and whoever was in charge of those supplies would be held responsible.

[2] Augustus M. Erwin of Rome.

the p.m.

Mail came at 7:45 p.m. Received two letters from Hannah. Immediately answered them. At 10:30 p.m. the regt as "fallen in" & that statement of Genl Grants captures of prisoners and guns made to the men. A continued cheer was the result. For half an hour the hooting was continued, at which time the men became quieted. The men as they came into line were in many cases naked except a shirt. Many were half dressed. A majority were without full suits. A grotesque spectacle to be seen by the light of the moon. The naked legs were a strong contrast to the sky blue of those who had their Trowsers on. Our Army is being successful.

Capt. Frank Lay, Company D.
*Oneida County History Center.*

April 7th Friday. Cloudy with light intermitting showers. Drills were some what lax to day. Finished my pants & had them on. The alterations improve them very much. The Chaplain brot the statement to camp this p.m. that Lee had surrendered to Grant. The men were full of yell. But we have recd no orders regarding the subject. It may be a canard. Some of our Recruits are making themselves agreeable in singing at Capt. Lay's[1] Quarters. Wrote again to Hannah. Boys all in good spirits.

April 8th Saturday. Breezy and warm. Camp duties were attended as usual. Drills were a deal more strict. Col. R. Daggett conducted the Battalion in the p.m. The men did well. Col. Daggett is as much at home in Drill as ever. Orders for moving were read to day, on Evening Dress Parade. We are to be in readiness on the morn of the 10th. We have had a good rest & shall be able for a long march. The men have received ample supplies of all needed articles of clothing, & Camp and Garrison Equipage, and will go forward with a replenished outfit. Wrote to John Rhodes. Some of our Officers have been over weening in their attachments for canteens filled with "B" whiskey.

1 Frank H. Lay of Utica.

April 9th Sunday. Last night the Officers at Hd Qrs had a Hugh drunken frolic. Officers from the 3d & 142d N.Y.V. came bringing the whiskey. Was kept awake till midnight. Day very fine. Preaching at 2:30 p.m. At night after getting to bed, the Q. M. came round and gave orders to pack up all things for which he was to furnish. Kendall & I packed our desk, & my knapsack, then went to bed again. Wrote to Hannah.

April 10th Monday. All ready for moving at 6 a.m. At 7:30 marched till near 2 p.m. & then were allowed 20 minutes for dinner, again starting on the tramp again. Halted in a considerably extended rolling field & formed Camp for night. Soon had tent up and supper cooked for eating.

April 11th Tuesday. Ready for moving at 6 a.m. Moved at 6:20 a.m. Marched till 10:20 a.m. and then halted for dinner. Started forward again at 2:30 p.m. This halt was at Bentonville which ville included a dozen wood colored buildings, mostly log. Went into camp about sundown. Had supper and went to bed.

April 12th Wednesday. Ready for moving at 6 a.m. But waited for trains to move, as our Brig was to act as rear Guard. While waiting Official news (orders) came that Lee had surrendered to Grant. A universal cheer & continued howling was immediately commenced. Caps, coats & even knapsacks were thrown into the air. Enthusiasm was the order of every man's feelings. After this we moved forward about a mile and then were forced to halt to enable the trains to pass a very bad "run" which crossed the road. The water in one portion of the run was so deep that the wagons would go down half over the boxes. Horses & mules would go in up to their shoulders. At this deepest point was a very bad pitch hole. A number of teams got stuck. Some of the animals fell down at this point. Only one team was permitted to pass this bad place at once. As soon as that train had got beyond it, another was ordered forward. At 6:15 p.m. our Brig had got over this, passing by on rails & timbers placed on stumps & roots on either side of the road. At 6:30 we took up line of march in rear of the trains, plodded along till 12 midnight. No signs of camp. Moved forward. Saw fires in advance. Thought they were camp fires. Began to think of going to rest. March on. Found we were deceived, it was the woods on fire. Kept moving. Men now were swearing & cursing every General in the column. Moved on, other fires were again deceived. Men are swearing terrificly. At 20 minutes of 4 a.m. we caught up with the column, filed into a field, formed regimental line, stacked arms, laid down, & sunk to sleep. We made about 15 miles during the night.

April 13th Thursday. At 5:30 a.m. Reville was beaten. After we had had about 2 hours sleep we were called to get ready for march. Made breakfast, and at 7 a.m. moved forward, our Brig taking the lead. Moved onward. The country is considerably rolling. Many Granite boulders are met with, especially beside the streams. At 9 a.m. came to a halt where two roads converged; in the one that leads to Raleigh, N.C. our column was moving on the left road. The 23rd Corps was on the other & having reached this point first we were compeled to halt till they had passed. After the 23rd A.C. had passed, we moved forward. Halted at 6 p.m. Went into camp in an open field. In the locality of this halt, Kilpatrick[1] had a brush with the Johnnies & made them dust at a rapid rate. A few horses were laying about dead. But I saw no graves. Not many could have been injured. The soil has become considerably stony. A coarse quality of quartz & silex stone are the general portions of the stones.

April 14th Friday. Fair. Ready for moving at 8 a.m. Marching to day was very regular. Soil more rolling & more cultivated as we neared Raleigh which place is 14 miles from our starting point of this mornings. Houses & out buildings are better, being mostly of board instead of logs. At 3:17 p.m. I caught sight of a steeple in R[aleigh]. A mile farther on & we got within sight of the extended, rolling slope of cleared land which bounds R. on this side. On this slope is encamped the 20th & 23rd A.Cs. Beyond them is a skirt of forest which hides R. from our view. A few steeples & domes are all we can see of R. from this point. We went into camp on the left of the R. Road, the other corps being on the right. Brass bands made music during the eve. Went to bed early & laid & listened to the music.

April 15th Saturday. Rain fell during the night. Thunder shower at morn. Orders to be ready & move at 9. Rain was falling, men began to get their tents down. Orders to move were countermanded. A general howl followed. News came that Johnson had surrendered to Kilpatrick, a continued Cheer & howl. Refixed tents & got in out of the rain. Remained in camp all day. At eve walked with Kendall as far as the State Capital in Raleigh. It was dark. Will not write of it now, wait for future visit.

April 16th Sunday. Warm Sun & bracing air. Earth drying fast under influence of wind & sun. Kendall & myself went to R[aleigh] to day. Attended church on the west side of the Capitol square. The Church is

---

[1] Brig. Gen. Judson Kilpatrick commanded Sherman's cavalry.

The capital building in Raleigh. *Library of Congress.*

just opposite the western entrance to the Capitol. Baptist Church. A brick structure plastered to represent Brown Sand Stone. Fine structure. Nearly finished in the interior. The ceiling is concaved with the slope of the roofing. The Preacher was a fine appearing man about 35 years of age. Thorough Education, a superior speaker, and a sound thinker. Text "It is a faithful saying and worthy of all acceptation, that Christ died for sinners, of whom I am chief." A few of the usual attendants were present, but the Soldiers filled the church so early that few only could get seats. The Soldier audience were very attentive. After service Kendall & I walked about the city some extensively. Very little grading has been done. Where streets are level, it is mostly the work of nature. The walks have curb stones, & are elevated about a foot above the street level. The earth is hard, & in dry weather is as firm walking as on pavement. Some pavements of brick are met. The st[reet]s are seldom aided by coble or other paving. Streets are mostly at right angles, & parallel. Massive oak trees are seen in every direction. Elm is intersperced somewhat, while numberless other kind are growing in profusion. Ornamental shrubbery is employed for the embellishing yards and courts. This gives the city a very rustic appearance which is heightened by the absence of general paving. The residences are universally in good repair. Many of fine finish & structure. Exceeding few of the miserable shanties that have marked other cities. The Capitol is a Splendid Structure of gray Sand Stone, Three Stories high. Fronts East & West. Each Entrance is fronted by 6 fluted sand stone pillars. The yard or Court surrounding is a fine grass plot. Noble oaks are growing luxuriantly in every part. Raleigh is a very quiet city. The Deaf and Dumb asylum is also a noticeable structure. The Insane Asylum is a brick edifice 2 ½ stories, long, with a pillard entrance. It has an insignificant observatory on its top. Long & low are the ideas that fix in the mind. The windows have white trimming & green blinds. This structure is s. west of the city on an eminence. The R. R. depot was of brick with tin roof. Has been burnt by rebel hatefulness. Many citizens remain. The people have settled into the idea the Secession is a Humbug & is finished. They appear far from uncomfortable under the yoke of Yankee authority. Our Soldiers are courteous & well behaved every where, though they are very numerous. The whole of Shermans army is encamped about Raleigh.

The Hills are covered with camps or parks of trains or artillery. Our men are excited with victory an[d] prospects of Peace, & are at Every sort of sport. The R. R. as it passes here is a down grade from Raleigh. Platform Cars, hand cars, & small truck [?] cars are used by whose who like the sport, in ridding down hill. The cars are pushed up, & the men ride down. Some ??? of these cars has created some laughing during the day. Two platform cars ran against each other to day at eve, & were left. A train has just come up from Goldsboro. Some one halloed, attracting the attention of the engineer. The train was slackened so that a smart bunt [?] was all the concussion that resulted. The train is not moving yet. The two platforms are in the way. Lt. Adams[1] has just come to the regt. He was taken prisoner Oct 27/64. Has been home on leave of absence. Was paroled some time ago. Now is exchanged. He relates the same inhuman treatment that all others have iterated before. The Spirit of Rebellion is fiendish.

April 17th Monday. Beautiful spring weather. Drilling is commenced again. Sherman & Johns[t]on are considering the subject of the latters surrendering. Nothing of interest transpires. Went into town with Commissary Sergt. Wicks.[2] Saw Genl Sherman as he got off the cars upon coming from Johns[t]on's Army. He is an athletic strong man, about 5 feet 10 or 11 inches in height, regular features, firm expression. Brown whiskers & mustache, cut short. He had but a small accompaniment of guard & Staff. An order was recd to day communicating the fact of the assassination of the President. A universal gloom pervades the Soldiers on this news. A wish is felt by them that Johns[t]on will not surrender

---

[1] Henry L. Adams, a teacher from Vienna.

[2] John Bartlett Wicks, a farmer from Paris Hill.

so that this Army may vent its desire to revenge the assassination of Lincoln by going through Johns[t]on's Army. This is a most diabolical climax to the Emination from Hell called Secession. Wrote to Hannah. Had a short sing at eve. Green, Kendall & myself enjoyed it very much.

April 18th Tuesday. Sun, coolish fresh air. Guard mounting at 8 o'clock a.m. Camp Guard Employs 24 men 3 corps. 1 Sergt and an Officer of the Guard. Battalion Drill at 3:30 p.m. A Heavy Thunder Shower during evening. Mail came, bringing me two letters from my wife. They were warming to my sympathies & charming to my mind.

April 19th Wednesday. Fair with some cloud. Guard mounting at 8 a.m. Monthly Inspection at 9 a.m. made by Lt. Ross A. A. I. G. 1st Brig. Cooked bacon & fried hard tack for breakfast. Had beans for supper last night. Some nice fresh beef has been issued to day, we had our fill at noon & night. Have been busy Drilling for tomorrow. A Review of the 10th A.C. is intended. Two trains came from Goldsboro to night. They followed each other closely. Wrote to my wife to night. Some thunder in the distance, but no rain or storm has reached us.

April 20th Thursday. A ??? spatter of rain at morn. At 7 a.m. we began to get into shape for Review. The Regt was formed into 7 companies of 20 files each. Each company subdivided into Platoons. At 8 moved toward the City. At 8 ½ formed column by Platoons, moved into close order, & halted stacking arms. Rested here awhile. At 9:15 Commenced moving up Fayetteville St. toward the Capitol. Head of column to right, to left, to left again taking us round a block & on the street passing the Capitol. Genl Sherman sat on his horse inside the Capitol grounds. The men march in excellent order, keeping regular lines & making good wheels. Our Brig Band did itself credit in the music line. General Sherman returned the salute of each line officer. But when our ragged colors passed and were lowered in salutation the General took his cap from his head, holding it by his side till the color co[mpany] had fully passed. We did ourselves full credit on the Review to day. We march in Platoon through a number of the streets, & then facing right, filing left came into rout[e] column & passed into camp. Clouds gather & thunder mutters at 4 p.m. Nearly all the Generals of this large Army were present to witness the Review. The Capitol grounds were full. Our Regt being in lead of the Brig was in front of the column. Our Brig arrived at camp again at a little after 11 a.m. Dress Parade at 6 p.m. A Brass band from the 24th Wisconsin Regt was over & serenaded the Chaplain. They are Reported to be all Welsh. Wrote to Father in evening. Trains came up from Goldsboro to day. Got mail in the a.m. None for me. Hope deferred.

April 21st Friday. Cloudy, Sultry, Thunder, Some rain in p.m. Sent out 45 men under command of Lt. Glazier at 3:30 this a.m. on a tour of reconnaissance. Usual camp duties. Sergt (orderly) Grant[1] came to the regt. He was taken Prisoner May 16th, 1864. Was exchanged some two mo[nth]s since. Formed Regt for Battalion Drill at 4 p.m., rain fell rapidly, Battalion was dismissed in a hurry. Squad Drill has been had to day. Recruits are being drilled regularly. They will form a fine Army for future work. Mexico may fell them.[2]

April 22nd Saturday. Stirring breeze. Commenced to Index the Regimental Descriptive book. Went to the City & witnessed the Reviewing of the 20th A. C. by Genl Sherman. The Troops marched splendidly. They had no knapsacks, haversacks nor canteens, a very sensible order of being reviewed. The men will go through with a Review with far more ease. Almost every Major Genl of the Dept of Mississippi were present. Also many Brigs[3] were on hand.

Genl Sherman was the least Pretentious in appearance of all the Generals. Men who have been hanging about Hospitals nursing during the Summer, are now coming back to the Regt. There is little danger now. Mail came. Got a letter from Hannah. Wrote an answer at eve.

April 23d Sunday. Cool fresh air from the west. Inspection at 9 a.m. Went to Raleigh, attended same church as last Sunday. Heard same man. Citizen more generally at church than last Sunday. A number of young ladies were present. They have the style of wearing Jockeys,[4] but have no such fancy trimming as I found while home wounded. A Union Blockade has made the difference. Had services in Regt. Chaplain Jones officiating. Pay Rolls came at eve. Commenced signing them.

---

1 H. Dwight Grant, a lumberman from Boonville.

2 He probably means "feel" and not "fell." Taking advantage of the Civil War, France attempted to recreate an empire in the Western Hemisphere by occupying México under Emperor Maximillian. At the end of the war the U.S. sent troops to the Mexican border and France withdrew. Maximillian remained, but was captured and executed by Mexican forces.

3 In this context, a reference to brigadier generals rather than brigades.

4 Perhaps a reference to the new fad of "bloomers," the trend among women of wearing pants under their skirts. An early meaning of the word "jockey" was a boy or fellow.

Kendall is quite sick, feverish. Mail came, no letter for me. Went to bed at 9 p.m. My digestives not feeling very calm.

April 24th Monday. Fair & warm during the middle of the day. Rearranged camp to day. Pay Rolls were laid aside for the day. Are being signed to night. Paymaster is expected in the morrow. Kendall is better to day. Boiled some Skoke berry tops for greens.[1] They taste some similar to Asparagus. Kendall liked them very much. During evening wrote to Minnie. This R. R. is very busy getting up supplies. A large Army eats much.

April 25th Tuesday. Beautiful weather. Orders to be ready for marching at 6 a.m. 26th inst. Carts are to be dispensed with and pack mules used instead. This worries the officers some as they have been accustomed to have their grub and kit carried in a Cart. A prevailing opinion is that we shall not move far if we go at all. Pay Master Tarr gave the Regt a supply of Green backs to day. I recd Pay as Sergt for the 4 mo[nth]s ending Dec. 31st 1864. Went to town today and bought a cap. The Regt was Inspected to day by the A. A. I. G of the 1st Brig 3d Div 10th A. C. Sutlers are running considerable business in Raleigh. Few other Traders are transacting business on acct of a want of good for the market. Kendall is still improving. I feel much better than yester morn. Platner [?] has been about collecting money. Sutlers love money as well as most humans. Mail came at evening bringing me a letter from Mother was glad to get it. Paid Capt. Lay & Lt. Glazier each $10 ($20) A. J. Kendall paid me $10.

April 26th Wednesday. Clear & breezy. The movement ordered for to day did not take place. As soon as orders were countermanded, nearly every man commenced refixing his quarters. Improvements were commenced all round. I am being troubled with Diarrhoe, making me averse to motion of any kind. Make poor business of eating. Drills & routine of camp duties resumed as before orders to move.

April 27th Thursday. As yesterday. Improvements are going on. Drills & camp duty are in full blast. Terms of Johns[t]on's surrender is coming in.

---

[1] A form of pokeweed that is poisonous, but whose shoots and leaves are edible when cooked.

April 28th Friday. Clouds, with some spatters of rain. Was excused from duty to day. Laid very quiet and improved some. F. White[1] commenced to cook for us. Mail came. I recd a pack of envelopes from wife.

April 29th Saturday. Pleasant, rain after going to bed at eve. F. White did his best to get articles we wanted, but as all stores in the city were close[d] in mourning for our beloved, but dead President, he was unable to procure the desired articles. A cannon boomed out every half hour from Sunrise to its setting when a National salute of 36 guns was fire[d]. Thus ended the formal expression of our unutterable & heart felt sorrow for the Great & Good Dead.

April 30th Sunday. Fair, with clouds. Muster & Inspection at 9 a.m. Being excused I was not present. White got some skoke greens to day, and they relished nicely. They are slightly astringent. Chaplain Jones gave us a short talk at 2:30 p.m. He has no power to draw out the men. They have no interest in the man & many have none in the subject although a large number have more in the subject than in the man. Dress Parade at 5:30 p.m. The 20th A. C. left for Richmond to day. They expect 15 days march, 10 miles per diem. Wrote to Father to day. Mail came: nothing for me. Made Duplicate copy of Roster of com [missioned] officers during evening.

May 1st Monday. Pleasant. Was excused to day, but went after a train to bring some boards for fixing our house. Kendall went with the train for the purpose of over seeing the job. They succeeded in getting a prime lot.

May 2d Tuesday. Pleasant. Commenced building up a new house. Sided up 4 ½ feet with boards, put up rafters & ridge pole. Drew fly over and boarding up the front, & closing the rear with Shelters had a nice room 10 x 9 for our abode. Commenced duty to day. Am better.

May 3d Wednesday. Beautiful. Completed our house. Made table. Wrote letter for Sharp.[2] Battalion Drill. Brig Dress Parade. Was out 2 hours at Battalion Drill. Mail came. One from Hannah. Evening was passed in writing to Hannah, one of my first class entertainments.

---

[1] Francis P. White, a farmer from Rome.

[2] Probably Stephen Sharp who was born in England and a farmer in Sangerfield who enlisted in Company D.

May 4th Thursday. A Slight Sprinkle at 4 p.m. Camp duties & drills. Parades & fatigue as yesterday. The Brig Parades are very imposing.

May 5th Friday. Splendid. Duties as usual. Furloughs came back approved to day. There were 19; Mail came. John got 3 letters. Kendall & myself are about well again. We can eat with a relish. Frank White is doing well in getting up our meals. Had excellent baked beans for dinner. Warm biscuit for supper. Notification came to commanding officer that Commission had been forwarded to myself & those who were recommended at the same time. Had an agreeable sing to night.

May 6th Saturday. Hot (94°in Shade) windy. Went to town to day. Got Cap, Shirt, & Shave. Drills were discontinued to day. Brig Parade at evening. Dust flew greatly. Genl Terry[1] favored us with his august presence on the Parade & Review. Our regt did well. Wrote to my wife, during eve.

May 7th Sunday. Hot 95°. Attended church at Raleigh. Mail came bringing me two letters from Hannah. Immediately answered. Whiled the evening away by reading the entries in a confiscated Album. The Camp Equipage which was stored at Wilmington came up to day. Many of our men are getting back to the Regt. Our Chaplain had belly ache to day & could not, or did not preach. He is destitute of an abiding confidence of usefulness. He is aware that the men have no confidence in his pretentions. His acts have given his professions the most direct denial. Moonlight evenings.

May 8th Monday. Windy, a ???. Went to town to day. Wrote all the p.m. Made out Tri Monthly for Kendall. Did some trading.

May 9th Tuesday. Breezy & cool. Rain fell last night. The dust has lain quite close today. Battalion Drill at 2 p.m. Finished Index to Descriptive Book. It contains about 1500 names. John commenced to mess with us.

Kendall has been more indisposed for a few days, but feels more free from ill to night. Wrote to my wife at eve.

May 10th Wednesday. Warmer that yesterday. Had Brig Inspection, made by Brevet Maj. Genl Ames.[2] He was accompanied by his own

---

[1] Maj. Gen. Alfred Terry commanded the Tenth Corps including the 117th New York.

[2] Brevet Maj. Gen. Adelbert Ames was a division commander at the time.

staff, also field officers of other Brigs & Regts. They passed from left to right, looking very closely at the soldiers. Genl Ames Inspected the non. com. Staff, Color Guard, & rear rank of "H" Co. (1st in line). Each of the other co's. was Inspected by an officer designated by the Genl. In this manner all of the co's were inspected at once & were soon passed upon as to soldierly airs, cleanliness, & accoutrements decided upon, as were the arms. The Regt passed a very favorable Inspection. All had an air of satisfaction. Had a short sing at evening.

May 11th Thursday. Hot & Sultry. No Drills, Capt. Downer being away, aiding in the Inspection of the 3d Brig. Mail came at 4 p.m. I received a Commission as 2d Lt. from Governor Fenton. Also a letter from Fred Garrett. Thunder shower at 9 p.m. A detail of 2 Corpls & 20 men was called for at the time of the shower. My cloths got a little damp while getting it out. Made out required papers applying for Discharge, to enable me to Muster.[1]

May 12th Friday. Pleasant, but coolish. Wrote most of the day. Rain fell nearly all night. 27 Recruits came to regt. at 7 p.m. Drafted or Subs. Fine able men.

May 13th Saturday. Warm & beautiful. Had Battalion Drill at a.m. Wrote all afternoon. Walked up to the city after Supper. Wrote to wife during evening. Tis a pleasure thus to use the time.

May 14th Sunday. Beautiful. Co. Inspection after Guard mounting. Rode the Col. Horse to the City, & went to church. Attended the Episcopal. The officiating clergyman is an energetic nervous speaker. A sharp voice. Both he & the clerk wear specks. Saw some very fine appearing young ladies. Such sights do a soldier good, he being debarred from association with the fair skinned sex. But this deprivation is to be short, thanks to the passing months. Had services in Regt at 3 p.m. Lt. Col. Meyer & Lt. Fairbank[2] came back to Regt this morn. Wrote to Sister Mary.

May 15th Monday. Beautiful. The men had their tents down, & board ready to load by 8 a.m. when line was formed & the Brig moved to the

---

[1] Fairhead had to be discharged from his position as sergeant before he could be mustered in as an officer.

[2] John H. Fairbanks, a Rome cordwainer, had been wounded at Fort Fisher, January 15, 1865, as had Lt. Co.. Meyer.

locality of our new camp, a Splendid oak & Pine Grove. One load for each of the trains had gone before the Brigade moved. As soon as we had arrived at our Camping ground, the business of staking for the tents was commenced. I soon found the spot on which my tent was to have station. I commenced at once & by 3 p.m. had my tent up in exactly the same pattern of the old camp. I did the work all alone Kendall being sick. Many of the men have their tents up. But some whose lumber came late have not a tent in shape. One day more will see us with a fine camp beneath umbrageous[1] oaks. Then extensive policing will be the order. I was taken so short to day that my drawers caught the excrement. I soon slipped off my garments & cleansed myself in a stream nearby. My drawers were left as unclean. Lt. Col. Meyer assumed com'd of Regt to day.

May 16th Tuesday. Beautiful. Men in regt have been very busy during the day, building Qrs making improvements on those built yesterday. Policing has been very effectively practiced also. Our camp begins to look as clean as a Kitchen floor. Wrote most of p.m. & evening. Mail came. I recd a letter from my wife, also one from John Rhodes of South Trenton. Kendall is much worse, has considerable fever. He has done no duty today.

May 17th Wednesday. Capital weather. Kendall is a little better, but has considerable fever. Have been busy writing & detailing Fatigue squads. Spent the evening very agreeably in writing to my wife. Walked down town in the afternoon.

May 18th Thursday. Rain. Have been busy all day writing. Made out Tri Monthly Report to night. Mail came. I got a letter from my wife. John got one from sister Mary.

May 19th Friday. Heavy showers. Made some improvements on my shanty. Went to town, had an old negro cobbler fix the toes of my shoes, so that they would not as he termed it whet off at the toes. I would say stubb off at the points. He did it very neatly, & I paid 25¢, what he asked for the job. Major Bagg came last night. His wound is well, but the muscles are stiff. Had a splendid Strawberry short cake for supper. Paid 50¢ a q[uar]t for the berrys. Kendall is a little better. Had an exhibition of an armless man who used his feet as we would hands. Among other exploits, he strapped his razor & shaved negro who ac-

---

[1] Shady.

companied him & the man who officiated as chief ring master. He shot bow & arrow with much accuracy. He also wrote a very legible "foot writing." A better hand than one half of our men would. He no doubt had practiced considerable on the sentences he transcribed. I have seen one man before who was more expert than he with the feet. The exhibitor took a collection, & our boy gave some what freely. Some laughable remarks were made.

May 20th Saturday. Rainy. Was busy all p.m. writing. The men have been having much sport to day mustering some negroes & one or two of the Dead Beat sort of men in the Regt. The boisterousness of the laughing is in proportion to the height the one being mustered is thrown, or his fright.[1] One ??? of rough repute was mustered. Kendall shows few improving symptoms. But not being dangerous, we have no fears. He does no work at all.

May 21st Sunday. Fair & warm. Attended Epis[copa]l Church in the a.m. Our Chaplain gave us a talking to this p.m. The men have very little care to hear him. Had a very agreeable sing in evening. Green, Wade[2] & myself. Wrote at letter to my wife this morning, using the hour from 5 to 6 for the work. The Adjts little Pony slipped its halter last night & is not to be found as yet. Kendall is some better, judging from acts.

May 22d Monday. Pleasant. Drills were resumed to day. Wrote some. Walked into town this evening. Got some Ice Cream 50¢ a plate. Kendall is getting better again. He is able to do some writing. Adjt Roys was up to night. He is after his horse & truck. Mail came no letter for me.

May 23rd Tuesday. Exuberant. Have been busy most of the day in office. Drills as usual. Much fatigue has been done to day. Lieut. Adams Officer of the Day. Our Discharges for Promotion do not come round yet. We begin to think that we cannot be mustered. Care little. Wrote letter to Father.

May 24th Wednesday. Beautiful. Inspection at 10 o'clock a.m. Brig Parade at 6 p.m. Lt. Col. Meyer made a miserable bore of himself in

---

1 He is describing placing men in a blanket and throwing them high in the air.

2 There was another Wade in the regiment, but the reference is probably to Myron Waite, a Sangerfield farmer in Company D.

handling the Regt after Parade. Putty[1] men are a useless appendage to any sort of society or association. Every man in the Regt was thoroughly ashamed of the Regt during the jumble. Commands were giving wrong. Each Co. got detached irregularly, & in no lawful relation to each other. Bah! on such men. Wrote to Minnie.

May 25th Thursday. Same as 24th inst. Drills. Some stir was made this morn to try for a muster for us. No go. Boden's[2] discharge was on the way back. I got here about 1 p.m. and as 2 p.m. was the hour at which mustering was to terminate, He with the Adjt went to Keeler's office immediately. The Papers for the rest of us had been delayed longer than Boden's & the errors were corrected in Boden's before they were in ours. Hence the advantage to him. Our Commissions & papers returned to night respectfully referred to an extract from the A. G. O.[3] This disappointment is very provoking. To be disappointed just as a prize or object is within grasp is decidedly irritating to the human disposition. And such irritability is aggravated when the unfavorable development of circumstances is the result of a friends heedless mistakes. Yet under all this, it may be a train of incidents which will be for our advantage in the end, but in the passage they are distasteful. Did up the Glow worm this eve to send home. It is in a small bottle containing some decayed wood.

May 26th Friday. Thunder Showers all day. Wet & damp. A discharge of lightening struck a large oak just behind our camp, shivering it. The counter or action current caught one of our men making him helpless for a time, but he is much better & will get over it. The thunder which followed was terrific. A dog kept by one of the hos[t]lers dug out of camp yelping & with tail on a parallel with his back bone. Wet & nasty to night. Our linen roof makes a miserable cover.

May 27th Saturday. Cloudy & drizzley. No drills. Men are having much sport in mustering darks, and frequently one of the drone sort of men in the Regt has to be tossed from a blanket. The darkies who serve the officers are into it also, & every fresh one, or any one of themselves who shall do any thing to deserve it, has to take his turn. Every day has from 6 to 10 or 12 of these musters. It enliven the monotony of

---

1 Perhaps an inference that Meyer was not strong or commanding enough.

2 Frederick E. Boden, a butcher from Rome, placed the colors of the 117th New York atop the parapet at Fort Fisher, the first to be planted on the Rebel fortifications.

3 Adjutant General's Office.

this useless life.

Mail came to day, bringing me a letter from my deeply loved wife. I also see one from John H. French. Kendall has set to work to be discharged. I have made out what Papers were necessary for this work in its initiatory steps. He is very poorly & gets no better.

The Ponny of the adjt's which was lost a week ago, has been found & got back. It was secreted 12 miles from here.

May 28th Monday. Pleasant. Attended Baptist Church in morning. Our Chap. Played one of his ??? at preaching in the p.m. I went to the Methodist Church in the evening. The Minister was a rough ear splitting talker. Substance of his discourse very good but not marked by close thought. Kendall has not gone. His papers are in the red tape channel.

May 29th Monday. Warm. Kendall left us to day. He is considerably weak. The prospect of getting home infused some energy into him. H.B. Spooner[1] Co. "G" was detailed in his place as clerk. Drills & camp Fatigues as usual.

May 30th Tuesday. Sultry. 3d Brig was reviewed by Gen. Ames to day. Rumors of being mustered out come round through Head Qrs, orders are in the Department, and are expected to come to us. The men are much in haste to have them get about. Nothing is more distinctly apparent than that the Soldiers have no spirit of aggression. Now that the authority of the Government is fully established, all desire for war has vanished. The men are satisfied & universally want to get away from Military life. This gives a decided negative to the charge of Invasion against the Yankees, which was held by the Rebels. We are a peacefully inclined people, but must be loyal to our Government. There is a tremendous strength when imposition upon free principles causes the Strength to be exerted. The <u>rights</u> of the people are the principles which form the base of our Republican Institutions.

May 31st Wednesday. Fine weather. Considerable agitation about orders to muster out all whose time expires before the 30th of Sept. Was busy. Weekly Inspection was made in haste. Largish mail came. One or two for me. Mother & Hannah. Wrote to Hannah in Evening.

---

[1] Horatio B. Spooner was new to the regiment, enlisting at Albany on March 6, 1865.

June 1st Thursday. Beautiful. Being the day of "Humiliation & Mourning," in compliance with the Proclamation all drills were dispensed. The Review pos[t]poned. At 6 p.m. Divine Services were held under direction of Chaplain Hyde 112th N.Y.V. The 4 Regts of the Brig were present, the men with Side Arms. The regts formed a Hollow Square, closed in Mass. Hyde refered very appropriately to the Death of our late, & now highly honored President. A people might well mourn when such a noble member of our human family is laid away in the tomb. The Brig band was on hand to give us some music. Pens are busy. The men are highly exhilarated on account of prospects of going home. Every yell brings out a protracted howl. Those who are to remain feel a little cool.

June 2d Friday. Very hot. Drills all suspended till further orders. All attention is being fixed on the Muster Out rolls. Review came off at 3 p.m. The Brig did itself credit & our regt did credit to the Brig. We feel a just & creditable pride in good appearance.

June 3d Saturday. Hot, distant thunder during p.m. growing nearer at 5 p.m. The officers were examined this afternoon in compliance with orders from War Dept. Only Capt. Downer & Maj. Bagg express a desire to remain in the army: busy on Rolls. Much irritable work about getting out the first copy.

June 4th Sunday. Hot. Attended church M. E.[1] In the p.m. I attended Sunday School Singing School. It sounded good, though the singing was inferior. Evening attend Services in same church.

June 5th Monday. Hot. Every writer in the regt is busy at the muster out papers. Keeler is crowding.

---

*Fairhead's diary ends abruptly at this point, most likely because he was so busy completing all the muster rolls and other paperwork necessary before the regiment could head for home. Four day later, on June 9, the men boarded cars on the Raleigh & Gaston Railroad to begin their journey back to Oneida County. Reaching City Point on the 14th, they went aboard the transport* Northerner *to Fortress Monroe, and there transferred to the steamer* Edward Everett *for the journey to*

---

[1] Methodist Episcopal.

*New York City. On arrival in their home state, the men continued on to Albany by train, arriving on the morning of the 18th where they enjoyed a meal prepared by the local citizens before continuing by rail to Utica.*

*Thousands of people turned out to welcome survivors home with a brass band, speeches, and an enormous buffet dinner that was probably the most rewarding part of the reception next to being united with their loved ones. Only one final duty remained. The regiment mustered for the final time in Syracuse on June 25 to turn in their muskets and other equipment and to received their final pay from the government. They were now once again free to pursue whatever interests they may have.*

*Of the total wartime enlistment of 1,441, only 315 responded to the final roll call in Syracuse. Some were in hospitals, some had been discharged for disabilities, late arrivals were transferred to the 48th New York, and many lay dead. Regardless of their status, every veteran who served in the 117th New York Volunteer Infantry earned for himself both the honor or serving his country and the satisfaction of knowing that his sacrifices contributed to saving that nation and eliminating the scourge of slavery.*

The torn and tattered remnant of the colors given to the 117th New York by the ladies of Utica in 1862 testify to the arduous service the regiment had seen.
*Oneida County History Center.*

# Bibliography


## Documentary Sources

Davis, Alan P. Letter, President of Rochester Historical Society to Geoffrey C. Clifford, Curator, Oneida County Historical Society, April 5 and May 2, 1966, Manuscript Box 4, 1986, MSS.1, FAI.1, OCHS.

Fairhead, George. Annotated roster of the 117th New York Infantry, Oneida County History Center, Utica, NY.

__________. "Fifty Years Married," unidentified newspaper clipping, Oneida County History Center.

__________. Folder, miscellaneous papers, Oneida County History Center, Utica, NY.

Foote, John B. Letter to sister, July 6, 1864, John B. Foote Correspondence, Duke University.

*Gladiator*. Ship manifest, Ellis Island ship manifest collection.

"Nemo." Letter to the *Utica Morning Herald*, May 24, 1864.

117th New York Infantry, Regimental Descriptive Book Companies A-K, RG 94, National Archives and Records Administration.

*Sheridan*. Ship manifest, Ellis Island ship manifest collection.

## Published Sources

Culver, Milton H. *Sparks from the Camp Fire of the 117th N.Y. Volunteers, from 1862 to 1865, Relating a Few Historical, Truthful, Thrilling and Mirthful Incidents as Gleaned from the Diary of Sergeant M. H. Culver, of E Co.* (Rome: Guernsey, 1917).

Humphreys, Andrew A. *The Virginia Campaign of 1864 and 1865* (Edison, NJ: Castle Books, 2002),

Jackson, Harry F. and Thomas F. O'Donnell, eds. *Back Home in Oneida: Hermon Clarke and His Letters* (Syracuse: Syracuse University Press, 1965).

Longacre, Edward G. *Army of Amateurs: General Benjamin F. Butler and the Army of the James, 1863-1865* (Mechanicsburg, PA: Stackpole Books, 1997).

Mowris, James A. *A History of the One hundred and Seventeenth Regiment N.Y. Volunteers (Fourth Oneida), from the Date of its Organization, August, 1862, Till that of its Muster Out* (Hartford: Case, Lockwood & Co., 1866).

Phisterer, Frederick. *New York in the War of the Rebellion* (Albany: J. B. Lyon Company, 1912).

Thomas, Howard. *Boys in Blue from the Adirondack Foothills* (Prospect, NY: Prospect Books, 1960).

*Union Army: A History of Military Affairs in the Loyal States, 1861-65 – Records of the Regiments in the Union Army – Cyclopedia of Battles – Memoirs of Commanders and Soldiers* (Madison, WI: Federal Pub. Co., 1908), Vol. II.

*War of the Rebellion: A Compilation of the Official Records of the Union and Confederate Armies* (Washington, DC: U. S. Government Printing Office, 1880-1901).

# Index

33d ANNIVERSARY
117th N. Y. V.
New York Mills,
AUGUST 21, 1895.
Honorary.
40th ANNUAL
REUNION
117th N. Y. Vols.
SUMMIT PARK
1902

Made in United States
Orlando, FL
19 May 2026

81641882R10074